Ellsworth Douglass

Pharaoh's Broker

Being the very remarkable experiences in another world of Isidor Werner. Written by himself.

Ellsworth Douglass

Pharaoh's Broker

Being the very remarkable experiences in another world of Isidor Werner. Written by himself.

ISBN/EAN: 9783337393663

Printed in Europe, USA, Canada, Australia, Japan

Cover: Foto ©berggeist007 / pixelio.de

More available books at **www.hansebooks.com**

PHARAOH'S BROKER

BEING THE VERY REMARKABLE EX-
PERIENCES IN ANOTHER WORLD
OF ISIDOR WERNER

(WRITTEN BY HIMSELF)

EDITED, ARRANGED, AND WITH AN INTRODUCTION
BY ELLSWORTH DOUGLASS

LONDON
C. ARTHUR PEARSON LIMITED
HENRIETTA STREET W.C
1899

CONTENTS

		PAGE
INTRODUCTION: ELUSIVE TRUTH	. .	7

BOOK I. SECRETS OF SPACE

CHAPTER
I.	DR. HERMANN ANDERWELT . . .	19
II.	THE GRAVITY PROJECTILE . .	27
III.	STRUCTURE OF THE PROJECTILE . .	37
IV.	WHAT IS ON MARS?	48
V.	FINAL PREPARATIONS . .	57
VI.	FAREWELL TO EARTH	67
VII.	THE TERRORS OF LIGHT . .	81
VIII.	THE VALLEY OF THE SHADOW . . .	91
IX.	TRICKS OF REFRACTION . . .	99
X.	THE TWILIGHT OF SPACE . . .	108
XI.	TELLING THE TIME BY GEOGRAPHY . .	117
XII.	SPACE FEVER	126
XIII.	THE MYSTERY OF A MINUS WEIGHT . .	141

CONTENTS

BOOK II. OTHER WORLD LIFE

CHAPTER		PAGE
I.	WHY MARS GIVES A RED LIGHT	157
II.	THE TERROR BIRDS	170
III.	TWO OF US AGAINST THE ARMIES OF MARS	182
IV.	THE STRANGE BRAVERY OF MISS BLANK	192
V.	ZAPHNATH, RULER OF THE KEMI	204
VI.	THE IRON MEN FROM THE BLUE STAR	220
VII.	PARALLEL PLANETARY LIFE	240
VIII.	A PLAGIARIST OF DREAMS	249
IX.	GETTING INTO THE CORNER	260
X.	HUMANITY ON PTAH	275
XI.	REVOLUTIONIST AND EAVESDROPPER	283
XII.	THE DOCTOR DISAPPEARS	292
XIII.	THE REVELATION OF HOTEP	304

INTRODUCTION

Elusive Truth

IT was the Chicago *Tribune* of June 13th, 189–, which contained this paragraph under the head-line: "Big Broker Missing!"

"The friends of Isidor Werner, a young man prominent in Board of Trade circles, are much concerned about him, as he has not been seen for several days. He made his last appearance in the wheat pit as a heavy buyer Tuesday forenoon. That afternoon he left his office at Room 87 Board of Trade, and has not been seen since, nor can his whereabouts be learned. He is six feet two inches high, of athletic build, with black hair and moustache, a regular nose, and an unpronounced Jewish appearance. His age is hardly more than twenty-seven, but he has often made himself felt as a market force on the Board of Trade, where he was well thought of."

But it was the *Evening Post* of the same date which prided itself on unearthing the real sensation. A scare-head across the top of a first page column read:

"A Plunger's Last Plunge!"

"The daring young broker who held the whole wheat market in his hands a few months ago, amassing an independent fortune in three days, but losing most of it gamely on subsequent changes in the market, has made his last plunge. This time he has gone into the cold, kind bosom of Lake Michigan. Isidor Werner evened up his trades in the wheat market last Tuesday forenoon, and then applied for his balance-sheet at a higher clearing house! No trace of him or clue to his whereabouts was found, until the *Evening Post*, on the principle of setting one mystery to solve another, sent its representative to examine a strange steel rocket, discovered half-buried in the sands of Lake Michigan, near Berrien Springs, two days ago. Our reporter investigated this bullet-shaped contrivance and found an opening into it, and within he discovered a scrap of paper on which were written the words: 'Farewell to Earth for ever!' Werner's friends, when interviewed by the *Evening Post*, all positively identified the handwriting of this scrap as his chirography. It is supposed that he took an excursion steamer to St. Joseph, Michigan, last Tuesday or Wednesday afternoon, and walking down the shore toward Berrien Springs, finally threw himself into the Lake. Neither Israel Werner, with whom the dead man lived on Indiana Avenue, nor Patrick Flynn, the chief clerk at his office, can give any reason for the suicide, or explain the exact connection of the infernal machine (if such it be) with the sad circumstance. But they both positively identify the handwriting on the scrap of paper. We have wired our representative to bring the mysterious machine to Chicago; and those who think they may be able to throw any light upon the case, are invited to call at the office of the *Evening Post* and examine it."

The *Inter Ocean* developed a theory that the suicide was only a pretended one for the purpose of fraudulently collecting life insurance policies. It was cited that Isidor Werner had insured his life for more than $100,000, and this in spite of the fact that he had no family, parents, brothers or sisters to provide for; but had taken the policies in favour of his uncle, Israel Werner, and in case of his prior death, in favour of a cousin, Ruth Werner. This theory gained but little currency among those who knew the man best, and although the insurance companies prepared to resist payment of the policies to the bitter end, yet, as time went on, no one attempted to prove his death, nor to claim the handsome sum which would result from it. Moreover, Israel Werner and his daughter Ruth, the beneficiaries under the policies, persisted in believing that their relative was yet alive, though they could give no good reasons for so believing, nor explain his disappearance.

In its issue of June 15th the *Tribune* scouted the

idea of suicide altogether. It had a better and more plausible theory of the case. Isidor Werner had a large sum of money in the Corn Exchange Bank, drawing interest by the year. In case of either a premeditated or a pretended suicide he would most certainly have withdrawn, and made some disposition of, this money. In fact, he had, on the day of his disappearance, drawn out five thousand dollars of it in gold. For this coin the *Tribune* believed he had been murdered, and that they had a clue to the murderer. The vanished man had several times been seen in the company of a suspicious German, of intelligent but erratic appearance. This queer character lived in a hotbed of socialism on the West Side, and the young broker was supposed to be in his power. In fact, it was known for certain that the erratic German had secured a large sum of money from him, and that Werner had visited his rooms in the slums of the West Side more than once. Moreover, the two had made a secret railway journey together two days before the disappearance, and on the very day that Werner was last seen, the German had fled his lodgings without giving any explanation of his departure to his few acquaintances. When the *Tribune* reporter called at these lodgings, the landlord still had in his possession a gold eagle, with which the German had paid his rent, and in the grate of the deserted room were the charred remains of burnt papers. One of these was a rather firm, crisp cinder, and had been a blue-print of a drawing. As nearly as could be judged, from its shrivelled state, it appeared to be the plan of some infernal machine. The name of the fugitive was Anderwelt, and he called himself a doctor. Further investigations were being carried on by the *Tribune*, which promised to prove beyond a doubt that he was the murderer of Isidor Werner.

But the *Evening Post* still held the palm for sensations, and I copy verbatim from its columns of June 15th:

"It is rare that a newspaper, dealing strictly in facts, has to record anything so closely bordering on the supernatural and mysterious as that which we must now relate. The following facts, however, are vouched for by the entire editorial department of the *Evening Post*, and many of them by several hundred witnesses. We begin by apologising to the hundreds who have called at this office and have been unable to see the Werner infernal machine. We gave it that name in a thoughtless jest, but its subsequent actions have more than justified the title. Our reporter brought it from Berrien Springs, as directed, and deposited it in the court of the *Evening Post* building. As is quite generally known, this court is a central well in the building, affording ventilation and light to the interior offices, from every one of which can be seen what goes on in it. The well is spanned by a glass roof above the eighth storey. In this court, at eleven o'clock this morning, the entire editorial and a large part of the business staff of this paper repaired, to examine the mysterious rocket-like thing. A little lid was opened, showing the recess where the tell-tale scrap of paper, written by Werner, had been found. Inside there seemed to be a pair of peculiar battery cells, whose exact nature was hidden by the outer shell. Outside there were several thumb-screws, which were turned both ways without any apparent effect. While making this examination the machine had been set up on its lower end, and when it was again laid down it *refused to lie on its side*, but persisted in *standing erect of its own accord*. This was the more wonderful because the lower end was not flat, so that it would afford a good base, but was pointed. More than a hundred people saw it stand up on this sharp tip, saw it lift up light weights which were placed upon it to hold it on its side, and saw it quickly right itself when it was placed vertically but wrong end down.

"Thinking this queer property had been contributed to it in some way by loosening the thumb-screws, they were next all set down as tightly as possible, to see if this tendency to erectness would be lost. Then, to the astonishment of every one in the court, and of several hundred people who were by this time watching from the interior windows, this infernal machine, without any explosion, burning of gases, or any apparent force acting upon it, slowly *rose from the ground*, and then, travelling more swiftly, *shot through the roof of glass* and vanished from sight! Nor has the most diligent search enabled us to recover it. Does it possess the secret of Isidor Werner's death?"

But the Chicago *Herald* had been working thoroughly and saying little until its issue of June 16th, when it claimed the credit of solving the whole mystery. Its long article lies before me as I write: There had been no suicide; there had been no murder; there had been no infernal machine. Doctor Anderwelt was a learned man, and the warm personal friend of Isidor Werner. Both men had shared the same fate; they might yet be alive, but they were certainly *at the bottom of Lake Michigan together*! They were imprisoned there in a sunken submarine boat, which was the invention of Doctor Anderwelt, and was built with funds furnished by the young broker. The foundryman who had constructed the big torpedo-shaped contrivance had been interviewed. He knew both men, and they were on the most friendly terms. In a moment of confidence Doctor Anderwelt had told him the machine was for submarine exploration; had explained the four-winged rudder, which would make it dive into the water, rise to the surface, or direct it to right or to left. Moreover, there were closed living compartments, around which were chambers containing a supply of air. He himself had pumped them full of compressed air, and it was so arranged that foul air could be let out when used and new air admitted. When all had been finished the foundryman had shipped the new invention, *via* the Michigan Southern Railway, to the shore of the Lake near Whiting, Indiana. Next the *Herald* had sought and found the conductor whose train had hauled it to Whiting. He remembered switching off the flat-car there, and he was surprised on his return trip next morning to see the heavy thing already unloaded and gone.

Undoubtedly, the two men had made an experiment with the diving boat under the surface of the water; and its failure to operate as hoped had resulted in its sinking to the bottom, with the two men imprisoned in it. On no other hypothesis could its disappearance, and that of the two men, be so plausibly accounted for. But as they had stores of air, and probably of food, there was a possibility that they were still alive inside the thing in the bottom of the Lake! Only three days had elapsed since it had been launched, and the *Herald* was willing to head a subscription to drag the Lake and send divers to search for and rescue the two unfortunate men!

All this serves to illustrate the untiring energy of newspaper investigation, as well as the remarkable fertility of journalistic imagination; for none of these clever theories hit at the real truth, or explained the correct bearing of the astonishing facts which the newspapers had so industriously unearthed.

And if the mystery of the disappearance of Isidor Werner was uncommonly deep and wonderful, the explanation and final solution of it is not less marvellous. After a delay of more than six years, it has just now come into my hands whole and perfect. It is in no less satisfactory form than a complete manuscript written by the very hand of Isidor Werner! I came strangely into possession of it, and it relates a story of interest and wonder, compared with which the mystery of his disappearance pales into insignificance. But the reader may judge for himself, for here follows the story exactly as he wrote it. Upon his manuscript I have bestowed hardly more than a proof-reader's technical revision.

ELLSWORTH DOUGLASS.

Boston, U.S.A.,
December 13th, 1898.

BOOK I

Secrets of Space

CHAPTER I

Dr. Hermann Anderwelt

I HAD been busy all day trying to swarm the bees and secure my honey. The previous day had been February 29th, a date which doesn't often happen, and which I had especial reason to remember, for it had been the most successful of my business career. I had made a long guess at the shaky condition of the great house of Slater, Bawker & Co., who had been heavy buyers of wheat. I had talked the market down, sold it down, hammered it down; and, true enough, what nobody else seemed to expect really happened. The big firm failed, the price of wheat went to smash in a panic of my mixing, and, as a result, I saw a profit of more than two hundred thousand dollars in the deal. But, in order to secure this snug sum, I still had to buy back the wheat I had sold at higher prices, and this I didn't find so easy. The crowd in the wheat pit had seen my hand, and were letting me play it alone against them all.

After the session I hurried to my office to get my overcoat and hat, having an engagement to lunch at the Club.

"If you please, Mr. Werner, there is a queer old gentleman in your private office who wishes to see you," said Flynn, my chief clerk.

"Ask him to call again to-morrow; I am in a great hurry to-day," I said, slipping on one sleeve of my overcoat as I started out.

"But he has been waiting in there since eleven o'clock, and said he very much wished to see you when you had plenty of time. He would not allow me to send on the floor for you during the session."

"Since eleven o'clock! Did he have his lunch and a novel sent up? Well, I can hardly run away from a man who has waited three and a half hours to see me;" and I entered my private office with my overcoat on.

Seated in my deep, leathern arm-chair was an elderly man, with rather long and bushy iron-grey hair, and an uneven grey beard. His head inclined forward, he breathed heavily, and was apparently fast asleep.

"You will pardon my awaking you, but I never do business asleep!" I ventured rather loudly.

Slowly the steel-blue eyes opened, and, without any start or discomposure, the old man answered,—

"And I—my most successful enterprises are developed in my dreams."

His features and his accent agreed in pronouncing

him German. He arose calmly, buttoned the lowest button of his worn frock-coat, and, instead of extending his hand to me, he poked it inside his coat, letting it hang heavily on the single button. It was a lazy but characteristic attitude. It tended to make his coat pouch and his shoulders droop. I remembered having seen it somewhere before.

"Mr. Werner, I have a matter of the deepest and vastest importance to unfold to you," he began, rather mysteriously, "for which I desire five hours of your unemployed time——"

"Five hours!" I interrupted. "You do not know me! That much is hard to find without running into the middle of the night, or into the middle of the day—which is worse for a busy man. I have just five minutes to spare this afternoon, which will be quite time enough to tell me who you are and why you have sought me."

"You do not know me because you do not expect to see me on this hemisphere," he continued. "Nor did I expect to find you a potent force in the commercial world, only three years after a literary and linguistic preparation for a scholarly career. Why, the *mädchens* of Heidelberg have hardly had time to forget your tall, athletic figure, or ceased wondering if you were really a Hebrew——"

"You seem to be altogether familiar with my history," I put in with a little heat. "Kindly enlighten me equally well as to your own."

"I gave you the pleasure of an additional year of residence at the University of Heidelberg not long ago," he answered.

"I do not know how that can be, for to my uncle I owe my entire education there."

"Perhaps an unappreciated trifle of it you owe to your instructors and lecturers. Do you forget that I refused to pass your examinations in physics, and kept you there a year longer?"

"You are not Doctor Anderwelt, then?"

"Hermann Anderwelt, Ph.D., at your service, sir," he replied somewhat proudly.

"But when and why did you leave your chair at Heidelberg?"

"It is to answer this that I ask the five hours," he said slowly.

"Oh, come now, doctor, you used to tell me more in a two-hour lecture than I could remember in a week," I answered, taking off my overcoat, and touching an electric button at my desk. My office boy entered.

"Teddy, have I had lunch to-day?" This was my favourite question on a busy day, and Teddy always answered it seriously.

"No, sir, you have an engagement to lunch at the Standard Club," he replied.

"Telephone to Gus at the Club that I can't come up to-day. Also send over to the Grand Pacific for a good lunch for two. Have some beer in it—real Münchner, and in *steins*," I directed, and then I reclined on a long leather lounge, and motioned to the doctor to have a chair. He declined, however, and walked slowly back and forth before me as he talked, keeping his right hand inside his coat, and with the left he occasionally ploughed up his heavy hair, as if to ventilate his brain.

"A year ago I gave up theoretical physics for applied physics; I resigned my chair at Heidelberg, and came to this progressive city. I brought with me a working model of the greatest invention of this inventive age. Yet it was then neither perfect in design nor complete in detail. But now I have hit on the plan that makes it practicable and certain of success. I need only a little money to build it, and the world will open its eyes!"

"But you must pardon me if instead of opening mine I shut them," I interrupted, seeing the point quickly, and losing no time in dodging. "I have no money to invest in patent rights; but still, you must stay to lunch with me."

Just here the doctor seemed to find it necessary to diverge from the orderly course of his lecture as he had prepared it, and interject a few impromptu observations.

"Events are difficult to forecast, but the capabilities of a youth are harder to divine. One educates his son in all the fine arts, and he turns out a founder of pig iron. One's nephew is apprenticed to a watchmaker, and in a few years, behold, he is a great barrister. Your uncle educated you thoroughly in the old Hebrew and Chaldee of the rabbis, and, lo! you are now the *ursa major* of the wheat market.

"Just now you are in the centre of the kaleidoscope of success. Slater, Bawker & Co. were there a month ago, but now they are only bits of broken glass in the bottom of the heap! And you? you are really a twisted bit of coloured glass like the rest, but you chance to be thrown to the middle. The mirrors of public opinion multiply your importance half a dozen times, and behold you are reflected into the whole picture. But the kaleidoscope turns, and the pieces of glass are shifted. Other broken chips now at the bottom of the heap will soon be filling the centre!

"Permit me to change my figure of speech. You are sweeping back the waves of the sea while the tide is falling, and the wide-mouthed public looks on, and whispers about that your broom makes all the waves obey, and drives them back at will. Just when you begin to believe it yourself the tide may turn, and neither brooms nor all the powers on earth can then sweep it back.

"Isidor Werner, you believe yourself rich; but your wealth is like molasses in a sieve. If you do not dip in your finger and taste the sweet occasionally, you will have nothing to show for your pains in the end. I shall ask you for but a taste of the sweet now, so that I may preserve a little of it against that day which may come, when the sieve will be bright and clean and empty again!"

There was a knock at the door.

"Come in!" I shouted. "Nothing but this lunch can save me from your eloquence. You have already ruined me in three similes!"

The waiter arranged a bountiful and tempting luncheon on a writing table. I commenced on it at once, but the doctor, though repeatedly urged, persistently refused. He took a long draught at a *stein* of Munich beer, and continued :—

"My invention proposes to navigate the air and the ether beyond, as well as the interplanetary spaces," he said impressively.

"Flying machine, eh?" I sneered, between bites of planked whitefish.

"Indeed no!" he growled, as if he detested this name. "My invention is not a machine but a projectile. It is not self-propelling, because if it depended upon its own propelling apparatus, it could not in thousands of years navigate the interplanetary spaces. It is a *gravity projectile*, and will travel at a rate of speed almost incalculable. It does not fly, but its manner of travelling is more nearly like falling."

I gave the doctor a quick searching look to see if I could discover any signs of incipient insanity. I met a firm, steady gaze; an earnest, convincing look. Somehow, I felt there was something real and true and wonderful about to come from the great scholar before me, and that I must hear it and hear it all; that I must lend a serious and thoughtful attention. My eyes were rivetted upon the doctor's for fully a minute in silence.

"Go on," I said at last; "I am all attention."

CHAPTER II

The Gravity Projectile

HERMANN ANDERWELT had probably suffered many disappointments and waited long for a hearing. Now he seemed to feel that his opportunity had come, for he continued with growing enthusiasm:—

"Hitherto all attempts at space travelling have been too timid or puerile. We have experimented at aerial navigation, as if the brief span of air were a step in the mighty distance which separates us from our sister planets. As well might steamboats have been invented to cross narrow streams, and never have ventured on the mighty ocean! We have tried to imitate the bird, the kite, and the balloon, and our experiments have failed, and always must, so long as we do not look farther and think deeper. Every Icarus who attempts to overcome the force of gravity, which conquers planets, and propel himself through the air by any sort of apparatus, will always finish the trip with a wiser but badly bruised head."

"Still, it has been freely predicted," I ventured,

"that this century will not close without the invention of a successful air-travelling machine."

"And I alone have hit upon the right plan, because I have not attempted to *struggle against* gravity, but have made use of it only for propelling my projectile!" exclaimed the doctor triumphantly.

"But wait!" I interposed. "Gravity acts only in one direction, and that is exactly opposite to the one you propose to travel."

"That brings me to the very important discovery I made in physics two years ago, upon which the whole success of the projectile rests. You will remember that, according to the text-books, very little is known about gravity except the laws of its action. What it is, and how it can be controlled or modified, have never been known. Electricity was as much a mystery fifty years ago, but we know all its attributes. We can make it, store it, control it, and use it for almost every necessity of life. The era of electricity is in full bloom, but the era of gravitational force is just budding."

"Can it be that we have as much to learn from gravity as electricity has taught us in the last half-century?" I exclaimed, as my eyes began to open.

"I believe it will teach us far more wonderful things, because it will take us to unknown worlds, while electricity has been confined to Earth. Its realm is the wide universe. It will show us what life there is on the planets. It will make us at home with the stars.

"What!" he continued in a sort of ecstasy. "Do you think all great discoveries are over, all wonderful inventions made? As well might a trembling child, elated with the success of its first feeble steps alone, suppose it had exhausted all the possibilities of life. We are but spelling over the big letters on the title page of the primary book of knowledge. There be other pages and grander chapters further on. There be greater volumes, and sweeter, more expressive tongues which man may learn some day.

"Has a reasoning Divinity created the heavens and peopled the myriad stars with thinking, capable beings, who must be perpetually isolated? Or may they not know each other some time? But shall we attempt to sail the vast heavens with a paper kite, or try to fly God's distances with the wings of fluttering birds? Nay; we must use God's engine for such a task. Has He tied the planets to the sun, and knitted the suns and their systems into one great universe obedient to a single law, with no possibility that we may use that law for intercommunication? With what wings do the planets fly around the sun, and the suns move through the heavens? With the wings of gravity! The same force for minute satellite or mighty sun. It is God's omnipotence applied to matter. Let us fly with that!"

"But will you permit me to suggest that we are soaring before the projectile is built?" I put in.

"Quite right. Let us come back to Earth, and return to facts. My studies in physics led me to believe that all natural forces—gravity, centrifugal force, and even capillary attraction—are, like electricity and magnetism, both positive and negative in their action. If they do not normally alternate between a positive and negative current, as electricity does, they can be made to do so. Gravity and capillary attraction, as we know them, always act positively; that is, they always *attract*. On the other hand, centrifugal force always acts negatively; that is, it always *repels*. But each of these forces, I believe, can temporarily be made to act opposite to its usual manner. I know this to be the case with gravity, for I have caused its positive and negative currents to alternate; that is, I have made it repel and then attract, and so on, at will, by changing the polarity of the body which it acts upon."

"Now that I remember it," I added, "our original ideas of magnetism were that it simply attracted. We knew the lodestone drew the steel, but only on better acquaintance did we learn of its alternating currents, attractive and repellant."

"I have positively demonstrated with my working model that I can reverse the force of gravity acting upon the model, and make it sail away into space. I will show you this whenever you like.

It is so arranged that the polarizing action ceases in three minutes, after which the positive current controls, and the model falls to the Earth again."

"But have you ever attempted a trip yet?" I inquired.

"Oh, no. The model was not built to carry me, but it has demonstrated all the important facts, and I now need ten thousand dollars to build one large enough to carry several persons, and to equip it with everything necessary to make a trip to one of the planets. With a man inside to control the currents, it will be far more easily managed than the experimental model has been."

"Suppose you had the projectile built, and everything was ready for a start," I said, "what would be the method of working it?"

"I should enter the forward compartment," began the doctor.

"But would you make the trial trip yourself?"

"I certainly would not trust the secret of operating the currents to any one else," he remarked, with emphasis. "And will you accompany me in the rear compartment?"

"No, indeed; unless you will promise to return in time for the following day's market," I replied.

"Then I shall engage some adventurous fellow as assistant. First, we must set the rudder, which is both horizontal and vertical, so that the projectile can be steered up, down, or to either side. Having fixed it so as to be directed a little upward,

I begin with the currents. Suppose the projectile weighs a ton, I gradually neutralize the positive current, which we are acquainted with as gravity. When it is exactly neutralized, the projectile weighs nothing, and the pressure of the air is enough to make it rise more rapidly than a balloon. When I have created a negative current, the projectile acquires a buoyancy equal to its previous weight. That is, it will now *fall up* as rapidly as it would previously have fallen down. It will not do to put on the full negative current at once, for we should acquire a velocity that would simply burn us up by friction with the atmosphere. However, the air is soon passed; if in the ether beyond there is very little friction, or none at all, we shall go at full speed, which will be the constantly increasing velocity of a falling body.

"Somewhere between the Earth and the nearest planet," he continued, "there is a place where the attraction of one is just equal to the attraction of the other; and if a body is stopped in that fatal spot it will be anchored there for ever, by the equally matched forces tugging in opposite directions. There is such a dead line between all the planets, and our principal danger lies in falling into one of these, for we should remain there a twinkling star throughout eternity! We must trust to our momentum to carry us past this point, and into space where the gravitational attraction of the other planet is paramount. Then we must promptly change our current from negative to positive, so that the other planet will attract us to her. Otherwise, she would repel us back to the dead line.

"With a positive current we are now literally falling into the new planet. We need not land unless we wish, for as soon as we enter a resisting atmosphere we can steer a course lacking barely a quarter of being directly away from the planet, just as you can sail a boat three quarters against the wind."

"But suppose you experiment at making a landing on this new planet?" I suggested.

"Very well. Of course, as soon as we enter an atmosphere, it behoves us to travel slowly to avoid overheating. We can still safely travel several hundred miles an hour, however. We continue falling until rather near the planet; then, turning the rudder gently down, we can sail around and around the planet until we choose our landing place. Gently reversing currents, a mild negative one soon overcomes our momentum. Tempering our currents experimentally to the pressure of the air, we can, if we desire, float like a feather and be wafted with every breeze. Just a suspicion of a positive current brings us gently to the surface, and, when we have cooled, we unscrew the rear port-hole and crawl out to explore a new world."

I had mentally made the trip, and was not only intensely interested, but infinitely pleased. I was

C

lost for some time with my imagination on the new sphere, but presently my mind returned to the practical side of the question, and I inquired,—

"Are you quite sure that ten thousand dollars will be sufficient to build and fully equip the projectile?"

"Yes, quite certain," he answered with decision. "It will be ample for that and for the expenses of forming a corporation to own my patents and exploit the invention. It is easy to see the projectile will be cheap of construction. No machinery is necessary; no strong building to withstand enormous shocks or anything of that kind. The principal expenditures will be for stores of food and for scientific and astronomical instruments. Of course, I wish to show you my working model and my plans for the practical projectile, and to explain to you many further details."

It was growing dark. I arose, turned on the electric light, and rang my bell. The office boy entered.

"Teddy, tell all the boys they may go, except Flynn. Ask him to wait, please."

I was quite used to making ten thousand dollar bargains in a few seconds of time and without the scratch of a pen. I had risked more money than that on the fact that Slater looked worried and Bawker was very cross when at his office, and it had won immensely. But here, what a prospect, what a far-reaching, all-encircling prospect it was! No time was to be lost; besides, there was pleasure to me in driving a good bargain and driving it quickly.

"And if I give you the ten thousand dollars, what do I get in return?" I asked, mentally placing my part at fifty-five per cent. of the shares at the lowest, so that I might control the company.

"You may organize the corporation yourself. The projectile must bear my name, and I must have the credit for all discoveries and inventions. Then you may give me such a part of the shares of the company as you think right," he replied.

On hearing this, I mentally advanced my portion to seventy-five per cent. Then I said,—

"When the projectile is built and proves successful, who is to manage the affairs of the company? Who is to finance it and raise further funds for exploiting its business?"

"I have no capacity for business," he declared. "I have no ambition to be a Pullman or an Edison. I would rather see myself a Franklin or a Fulton. You shall manage all the business affairs."

"Then I will undertake the whole matter, and give you my cheque for ten thousand dollars tonight, provided you allow me—ninety-five per cent. of the company's shares!" I said, simulating a burst of generosity.

Doctor Anderwelt ploughed his hair and harrowed his beard. He knew this was giving too much, but to have the projectile built, to sail away, to make all those grand new discoveries, to write books, and have future generations pronounce his name reverently along with Kepler and Newton! I did not believe he would have the courage to say no. While he meditated, my bell summoned Flynn.

"Please draw a cheque for ten thousand dollars to the order of Hermann Anderwelt," I said, watching the doctor as I spoke. There was indecision in his face.

"Suppose I allow you, say, ninety per cent.?" he said at last.

I was signing the cheque Flynn had brought me. "Done!" I cried, handing it over. He scanned it carefully, and after a long time said,—

"Mars is nearest to the Earth on the third day of next August. Fortunately Chicago is a good place to do things in a hurry. The projectile must be ready to start early in June, but its construction and its first trip must be kept a profound secret."

The doctor must have been hungry since noon. He began munching a chicken sandwich. The cold planked whitefish tasted quite as good as smoked herrings, perhaps, and strawberry short-cake in March was a luxury which he evidently appreciated.

CHAPTER III

Structure of the Projectile

A FEW weeks later I received a letter from Dr. Anderwelt asking me to call at his rooms on the West Side that afternoon, as soon as the market had closed. He desired to exhibit and explain the drawings of the new projectile and talk over the preparations for the trip. I had been so engrossed with every sort of worry that I had thought but little of the doctor and his grand schemes of late. But now I was anxious to know what progress he was making. Sometimes I felt that I had been foolish to put any money into the thing; but the doctor's idea of reversing gravity was so simple and so elemental, that I marvelled it had never occurred to scientists before.

After the market I hunted up the street and number the doctor had given me, and found a little, dingy boarding-house, lost among machine shops and implement factories, near the west side of the river. In a third-floor back room, with one small window looking out on dark, sooty buildings and belching chimneys, Dr. Anderwelt was thinking out all the incidental problems, and preparing for all the emergencies that might arise on a trip of some forty million miles, through unknown space, to a strange planet whose composition was unguessed.

The walls of the room were soiled and bare, except for blue-prints of drawings from which the projectile was being built in neighbouring foundries. There were but two plain, hard chairs in the room. The doctor sat on one with a pillow doubled up under him for a cushion. He was bending over a draughting board, which was propped up on the bed during the day and went under it at night.

Three flights of steep stairs had taken my breath, and I dropped into the other hard chair and exclaimed,—

"I say, Doctor, why didn't you take an office in the twelfth heaven of a modern office building over in town, where they have elevators? I have really forgotten how to climb stairs. Didn't I furnish you money enough to do this thing right?"

"Don't you think this is a good place?" he inquired in some surprise. "The rent is cheap, and it is convenient to the work. But speaking of elevators, we are going to revolutionize all that. No more hoisting or hydraulic lifts after we apply our ideas to the lifting of these elevator cages!"

"I am afraid this idea of negative gravity is apt to revolutionize everything, and generally upset the entire universe," I replied. "I have been wondering what would happen if you were to apply a negative current to this Earth of ours and send it whirling out of its orbit, an ostracised Pariah, repelled by all the celestial bodies!"

"Not the slightest danger of any such calamity," he answered. "The reversal of polarity can only be accomplished with comparatively small and insignificant masses. It would be impossible to impart a negative condition even to the smallest satellite. Our projectile will weigh but a few thousand pounds, compared to the millions of tons of the smallest celestial bodies. The Creator has looked out for the stability of the universe, never fear for that! And He has also given us a few hints of negative currents and repellant gravities in the form of meteorites and falling stars, which cannot be so well explained by any other theory. But what I want to talk to you about is the vital importance of providing against every possible emergency before starting on this trip through space. A trifling oversight in the preparations may mean death in the end, and things we put no value on here we might be willing to give a fortune for on Mars!"

"Well, let's hear how this thing is built," I said, rising and facing the larger blue-print. "So that's the shape of it, is it? Looks like a cigar!"

"Yes, the design resembles that of a torpedo considerably," replied the doctor, and referring to the sectional blue-print he began explaining the construction.

"This outer covering is a crust of graphite or black lead, inside which is a two-inch layer of asbestos. Both of these resist enormous heats, and they will prevent our burning by friction with atmospheres, and protect us against extremes of cold. Also, when we are ready, they will enable us to visit planets about whose cooled condition we are not certain. We might touch safely for a short time on a molten planet with this covering.

"Next comes the general outer framework of steel, just within which, and completely surrounding the living compartments, are the chambers for the storage of condensed air for use on the trip. These chambers are lined inside with another layer of asbestos. Now, air being a comparatively poor conductor of heat, and asbestos one of the best non-conductors we know of, this insures a stable temperature of the living compartments, regardless of the condition without, whether of extreme heat or extreme cold. Afterward comes the inner framework of steel, and lastly a wainscotting of hard wood to give the compartments a finish."

"How large are these living rooms?" I inquired.

"The rear one is four feet high and eight feet long. The forward one, designed for my own use, is longer, and must contain a good-size telescope and all my scientific instruments. The apparatus with which I produce the currents is built into the left wall, and it acts on the steel work of the projectile only. The rear compartment has a sideboard for preparing meals, which will have to be wholly of bread, biscuits, and various tinned vegetables and meats. We shall not attempt any cooking."

"But are there no windows for looking out?" I queried.

"Certainly, there are two of them, made of thick mica. One is directly in the front end, through which my telescope will look. The other is in the port-hole in the rear end. Each window is provided with an outer shutter of asbestos, which can be closed in case of great heat or cold. You will notice the two compartments can be separated by an air-tight plunger, fitting into the aperture between them. It will be necessary for both of us to occupy the same compartment while the air is being changed in the other. The foul air will be forced outside by a powerful pump until a partial vacuum is created. Then a certain measure of condensed air is emptied in, and expands until the barometer in that compartment indicates a proper pressure."

"The air will be made to order while you wait, then?" I put in.

"That is exactly what will be done in a more literal manner than you may suppose!" exclaimed the doctor. "This air problem is a most interesting one, for we must educate ourselves on the trip to use the sort of atmosphere we expect to find when we land. For instance, going to Mars we must use an atmosphere more and more rarefied each day, until gradually we become used to the thin air we expect to find there. Of course, there is an especially designed barometer and thermometer, capable of being read in the rear compartment, but exposed outside near the rudder. The barometer will give us the pressure of the earthly atmosphere as it becomes more and more rare with our ascent. It will show us what pressure there is of the ether, which may vary considerably, depending on our nearness to heavenly bodies. It will also immediately indicate to us when we are entering any new atmosphere. When we have arrived at Mars, we shall observe the exact pressure of the Martian air, and then manufacture one of the same pressure inside, and try breathing it before we venture out. The thermometer will give us the temperature of the ether, will indicate the loss of heat as we leave the sun, and will show us the Martian temperature before we venture into it."

"But you have said the condensed air will be used to resist the outer heat. This will certainly make it so hot it will be unfit to breathe," I interposed.

"Ah, but you forget that the quick expansion of a gaslike air produces cold. We shall regulate our temperature in that way. If it is becoming too warm inside, the new measure of condensed air will be quickly introduced into the partial vacuum, and its sudden expansion will produce great cold, and freeze ice for us if we wish it. On the other hand, if the compartments are already cold, we shall allow the condensed air to enter very gradually, and its slow expansion will produce but little cold. The question of heating the projectile is the most difficult one I have found. We cannot have any fires, for there is no way for the smoke to escape, and we cannot carry oxygen enough to keep them burning. I have decided that we must depend on the heat arising from outer friction and from absorption of the Sun's rays by our black surface. When we are in ether where friction is very little, the velocity will be all the greater, and I believe we shall always be warm enough. You must remember, we shall not have the slightest suspicion of a draught, and we must necessarily take along the warmest clothing for use on Mars. Even then we probably cannot safely visit any but his equatorial districts."

"This is the rudder, I suppose; but haven't you put it in wrong end first?" I asked. "It is just the opposite of a fish's tail. You have the widened end near the projectile and the narrow end extending."

"Yes, and with good reason. You will note that the rudder slides into the rear end of the projectile so that none of it extends out. This is a variable steering apparatus, adapted to every sort of atmosphere. Naturally, a rudder that would steer in the water, might not steer the same craft in the air. There is probably a vaster difference between air and ether than between water and air. It is necessary, therefore, to have a small rudder with but little extending surface in thick atmosphere; but when it becomes thinner the rudder must be pushed out, so that a greater surface will offer resistance. When we start, the smallest portion of this rudder moved but the sixteenth of an inch, up, down, or to either side, will quickly change our course correspondingly. When we have reached the ether, the full surface of the rudder pushed out and exposed broadside may not have much effect in changing our course. This is one of the things that we cannot possibly know till we try. However, if ether is anything at all but a name, if it is the thinnest, lightest conceivable gas, and we are rushing through it at a speed of a thousand miles a minute, our rudder certainly should have some effect."

"But suppose you cannot steer at all in the ether, what then?" I interposed, hunting all the trouble possible.

"Even that will not be so very dreadful, provided we have taken a true course for Mars while coming through the Earth's atmosphere. There is no other planet or star nearer to us than Mars when in opposition. Therefore there will be nothing to attract us out of our correct course; and if we can manage to come anywhere near the true course, the

STRUCTURE OF THE PROJECTILE 45

gravitational attraction of Mars will draw us to him in a straight line. The Moon might give us some trouble, and we shall be obliged, either to avoid her entirely by starting so as to cross her orbit when she is on the opposite side of the Earth, or else go directly to the Moon, land there, and make a new start. But if the ether which surrounds the Moon (for she has no atmosphere so far as we know) has no resisting power whatever, we might have rather a difficult time there. The only thing we could do would be to land on the side toward the Earth, then disembark and carry the projectile on our shoulders around the Moon to the opposite side, making a new start from there!"

"What on earth do you mean?" I exclaimed, interrupting. "Land on a satellite which has no atmosphere, and carry this projectile, weighing over a ton, halfway around the globe?"

"But the point is, it isn't on the Earth, but on the Moon! Think it over a little, and see how easily we could do it now. In the first place, we shall always carry divers' suits and helmets, to use in going ashore on planets having no atmosphere. Air will be furnished through tubes from inside the compartments. In the second place, the projectile in its natural state will hardly weigh two hundred pounds on the Moon, since the mass of that satellite is so much less than the Earth's, and weight therefore proportionately less. But you must remember I can make the projectile weigh nothing at all, so one of us could run ahead and tow it, as a child would play with its toy balloon."

"I perceive you have already made this trip several times, and are quite familiar with everything. But in case the Moon's surface is not suitable for foot passengers, what then? I understand it to be rough, jagged, mountainous, and even crossed by immense, yawning, unbridged fissures."

"That is most likely true, and for that reason we must carry a jointed punt-pole, and take turns standing on the back, landing and punting along through space just above the surface. Do you remember how far you can send a slightly shrunk toy balloon with one light blow? And how it finally stops with the resistance of the air? Without any resisting atmosphere, how far and how easily could it be sent along?"

"I can quite imagine you, astride the rudder of this thing, with a punt-pole as long as a ship's mast and as light as a broom-straw, bumping and skipping along in the utter darkness on the other side of the Moon; scaling mountains, bridging yawning chasms, and skimming over sombre sea-beds!" I laughed, for it aroused my active sense of the ridiculous.

"And the Moon may be well worth the exploration," exclaimed the always serious doctor. "Who knows what treasure of gold and silver, or other metals, rare and precious here, may not be found there? Why was the Moon ever created without an atmosphere, and therefore probably without the possibility of ever being inhabited? Is it put there only to illume our nights? Remember, we do the same service for her fourteen times as well; and if she has inhabitants they may think the Earth exists only for that purpose. Is it not more reasonable to suppose that some vast treasures are there, which the Earth will some day be in pressing need of? That it is a great warehouse of earthly necessities, which will be discovered just as they are being exhausted here? And who knows but *we* may be the discoverers ourselves? If the satellite is uninhabited, it will belong to the first explorers. Its treasures may be ours! We shall at least have a monopoly on the only known method of getting there and bringing them away."

"Ah! now you tempt me to go with you," I said, in a mild excitement. "Now I see myself, erect on the rudder, a new Count of Monte Cristo, waving the long punt-pole majestically, and exclaiming, '*The Moon is mine!*'"

CHAPTER IV

What is on Mars?

"I ONLY wish you *would* come along with me," replied the doctor. "I have no idea what intelligent, educated person I can persuade to accompany me, unless he is given an interest in the discoveries. You are the person most interested in the enterprise, and you should go. If it is money-making that detains you here, you may rest assured that we shall find fortunes for both of us somewhere."

"I am a slave to the excitement of my business," I answered. "I could not possibly spend two or three months in a lonely cell, flying through space, without a ticker or a quotation of the market. Besides, there are people on the earth I should not care to leave, unless I was certain of getting back soon."

"You may be sure of excitement enough, and of a continuously novel kind. Besides, of what interest are the people of this earth, who are all alike, and whom we have known all our lives, compared with the rapture of finding a new and different race, of investigating another civilization, and exploring an entire new world?"

"I shall have to warn my friends about you and have myself watched, lest you persuade me and run away with me when the time comes. If your adventures are half as exciting and varied as your theories, I should hate to miss them. But tell me why you have chosen Mars for a first visit."

"Because of all the planets he is the one which most resembles the Earth in all the essential conditions of life. He is the Earth's little brother, situated next farther out in the path from the Sun. He has the same seasons, day and night of the same length, and zones of about the same extent. He possesses air, water, and sufficient heat to make habitation by us quite possible. Moreover, his gravity problem will not put earthly visitors at a disadvantage, as it would on the very large planets, but rather at a distinct advantage over the Martians."

"What do you expect to find on Mars?" I queried.

"That is a very comprehensive question, and any answer is the merest guess-work, guided by a few known facts," replied the doctor. "The principal controlling fact is the reduced gravitational attraction of Mars, which will make things weigh about one-third as much as on the Earth. The air will be far less dense than here. In the mineral kingdom the dense metals will be very rare. I doubt if platinum will be found at all; gold and silver very little; iron, lead, and copper will be comparatively scarce, while aluminium may be the common and useful metal. Gases should abound, and doubtless many entirely new to us will be there. It is not unlikely that many of these will serve as foods for the animals and intelligent beings. It is also quite possible that the heavier gases may run in channels, like rivers, and be alive with winged fish and chameleons."

"How about vegetation?" I suggested.

"The vegetable kingdom will certainly not be rank and luxurious, because there is not enough sunlight or heat for that; nor will it be gnarled and tough, but more likely spongy and cactus-like. The weak gravity will oppose but a mild resistance to the activity and climbing propensities of vegetable sap, however, which is likely to result in very tall, slender trees. The forces that lie hidden in an acorn should be able to build a most grandly towering oak on Mars. Among the animals the species of upright, two-legged things is apt to abound. There is no reason for four legs when the body weighs but little. On the Earth an extremely strong development of the lower limbs is necessary for upright things, as is shown in the cases of kangaroos and men. In order that a cow might go comfortably on two legs, she would have to be furnished with the hind-legs of an elephant; but not so on Mars. Creeping things would be very few, and it is possible that fish may fly in the water with a short pair of wings. What four-legged animals there are will very likely be large and monstrous; for an enormous animal could exist comfortably and move about easily without clumsiness. For instance, an earthly elephant transferred to Mars would weigh only one-third as much, and so there might well be elephants three times as large as ours, perfectly able to handle themselves with ease."

"By the same reasoning then, I suppose the intelligent beings, or what we call men, will be great giants twenty-five feet high?" I put in.

"Some have thought so, but I do not at all agree with them," replied the doctor. "I stick to the theory of small men for small planets, and large men for large planets. There is no possible reason for a large man on Mars, where muscular development is uncalled for and useless, and where the inhabitable space is small. If there are men on Jupiter, they must of necessity be enormously strong to hold themselves up and resist gravitation. If they walk upright (which I think unlikely), their legs must be very large and as solid as iron. The Martian legs are likely to be small and puny, and I believe the upper limbs will be much more strongly developed. In fact, on Mars the Creator had His one great opportunity of making a *flying man*, and I do not think He has overlooked it. With a rather small, tightly-knit frame, and the upper limbs developed into wings as long as the body, flying against the weak Martian gravitation would be perfectly easy, and a vast advantage over walking."

"Ah! then perhaps they will fly out to meet you!" I ejaculated.

"If they do, they will be stricken with fear to see that we fly without wings and so much more rapidly," he answered, and continued: "If a flying race has been created there, we shall probably find the atmosphere deeper and relatively (though not actually) denser than the Earth's. This would serve to add buoyancy and still further diminish weight, thus making flying quite natural and simple. I certainly do not believe that the Martians are subjected to the tedium of walking. If they do not fly, they will at least make long, swift, graceful hops or jumps of some ten or fifteen feet each. This would require a more hinged development of the lower limbs, like a bird's. It is also possible that the lower limbs may have the prehensile function, and do all the handling and working."

"But how about intelligence and intellectual development? That is the main thing, after all," said I.

"To answer that takes one into the realm of pure speculation. There are but few facts to guide one's guesses. But the trip yonder is worth making, if only to learn that. I do not incline to the opinion that their civilization is vastly older and more developed than ours. Granting the nebular theory of the origin of the universe (which is, after all, only a guess), it is not even then certain that Mars was thrown off the central sun before the Earth. It is much smaller, and may have been thrown off later and travelled farther for this reason. Another good reason for believing in a less advanced civilization is the length of the Martian year and consequent sluggishness of the seasons. He requires 687 of our days to complete his sun revolution, making his years nearly twice as long as ours. I believe his whole development is at a correspondingly slow rate of speed."

"Which do you think is the most advanced and enlightened planet, then?" I ventured.

"That one which finds a way to visit the others first," he answered, with a touch of pride.

"But there may be a tinge of personal conceit in that idea," I suggested.

"Possibly a mere tinge, but the essence of it is apparent truth," he declared. "That planet which has learned the most, made the greatest discoveries and the most useful inventions, is the best and fittest teacher of the others, and will be the sharpest and keenest to gather new information and formulate new science. It is eminently fit that representatives of such a planet should visit the others, and eminently unfit that any primitive civilization engaged in base wars and striving for mere conquest should be allowed that privilege. An all-wise Creator would not permit a huge, strong, ignorant race entirely to overrun and extinguish one weaker but more intelligent. He might permit a strong, intelligent, masterful race to rule and direct a weaker and dependent one, as a schoolmaster rules and guides a child."

"Then you think we are the wise and masterful race?"

"As no other race has yet discovered us; as they have all left the Space Problem unsolved, and as it has been uncovered to us, that is my irresistible conclusion."

"Still, you will not go with ideas of conquest, but to teach and to learn?"

"We shall take with us swords, shields, and firearms, for defence. Unless I mistake the nature of their metals, our steel will resist any weapon they can manufacture. But what explosives or what noxious gases they may have, all strange to us, it is impossible to conjecture. Therefore, we shall go with peace in our hands."

"What progress do you think they have made in inventions?" I suggested, as the doctor hesitated.

"If they are winged men, I should say they have never felt that urgent need of railroads, steam boats, telegraphs and telephones, which was the mother of their invention here. Flying or air-travelling machines will no more have occurred to them than a walking machine to us. They will have thoroughly explored every part of their planet, and it is possible that their cities will have been built on high plateaus, or even on mountain peaks. But they will not have builded greatly, for they will have been able to use the great architecture of nature in a way impossible to us."

"Have you heard the theory advanced by some humorous scientist not long ago, that the organs of locomotion and prehension would some day, or on some planet, be supplanted by machinery, and that digestive apparatus would give way for artificially prepared blood?" I asked.

"Oh, yes, that fanciful idea is novel, but irrational. It makes man only a fraction of a being. On every planet, no matter what the advancement of civilization, we shall find *complete beings*, not dependent on adventitious machinery for locomotion or labour, or on artificial or animal blood for nutriment. Think how helpless such a creature would be at the loss or rusting of his machinery, and at the exhaustion of just the right sort of nutritive fluid. Our digestive apparatus will convert a thousand different foods into blood. Suppose we could live only on buffalo meat? We should all have been dead long ago. We might as well imagine men as mere fungus brains, swimming in rivers of blood; or as beings beyond the necessity of personal thought, and living on brain sandwiches, cut from the thinking heads of others. Eating is not only a necessity, but a pleasure——"

"That is just what I was thinking," I interposed, looking at my watch, for it was growing late.

"Well, now I have told you how I would have peopled Mars had the order been sent to me here to do it," said the doctor, "will you go along with me, and see how nearly I am right?"

"I am afraid not," I replied; "my business ties forbid. However, I want to see you make the start and the moment you return!"

CHAPTER V

Final Preparations

ON the tenth day of June, Dr. Anderwelt had written me as follows:

"Please catch the 7.25 train on the Lake Shore for Whiting this evening. I will take the same train, and we will walk from Whiting to a deserted railway siding two miles further on, where the projectile has been shipped. We will unload it from the flat car and take it into a grove of scrub oaks on the shore of Lake Michigan, near by. This will be enough to demonstrate to you our control of gravity. The experimental model is there also, and we will send it off on a trip if you like. Everything will be ready for the start to Mars to-morrow night."

I dined early and caught the train specified at Twenty-Second Street. The doctor was looking for me from the rear platform of a car. It was a local train, and crept slowly out through the smoky blackness of South Chicago, illuminated here and there by the flaming chimneys of her great iron furnaces, to the little city of pungent smells, of petroleum tanks and oil refineries, in Northern Indiana. The doctor was explaining the difficulties he had experienced in getting a companion for the trip.

"Men whom I could hire for mere wages are not intelligent enough to understand the workings of the projectile, or to comprehend the risks they may run. Besides, their companionship and assistance during the trip through space and on a new planet is worth nothing. On the other hand, I could not afford to go about explaining the workings of so important an invention miscellaneously to people capable of understanding it in an experimental search for a companion. I might not find one among twenty, and I would be tossing my secrets to the winds, and inviting all the daily papers to send their representatives to report the start. My reputation as a scientist, on the other side, is too dear to me to risk a public failure. If the projectile acts, as I am confident it must, on our return we shall take out letters patent and form our company to exploit the business features. But primarily, this is a test of the projectile and a journey of exploration and research. Business afterward."

Naturally on this point we had disagreed. My motto had always been "Business first!" and I had desired to have the patents secured immediately. But the doctor would not consent to the filing of the required specifications and claims, lest his secrets should be learned before success was demonstrated. As a compromise, the doctor had agreed to leave the necessary descriptions and data in a sealed envelope with me, which I was to be at liberty to open and place on record at any time during the doctor's absence that I might deem it necessary in order to protect our rights.

" Whom have you finally secured to go with you, then ? " I asked.

" I will tell you that after we have finished to-night's work," said the doctor, and then abruptly changed the subject.

The walk from Whiting was inspiriting. It was a beautiful night. There was not a cloud in the sky and no Moon, which made the stars all the brighter. Everything was still, save the constant lapping of the great lake on the sandy shore, but a short way off.

" Yonder is the mustard seed planted in the heavens, which shall grow into a whole new world for us ! " exclaimed the doctor, pointing out a particularly bright star. " That is Mars rushing on to opposition. In six weeks he will be nearest to the Earth ; so for that time he will be flying to meet us. To-morrow is our last day on Earth ; to-morrow night the ether ! And in six weeks, diminutive but mighty man will have known two worlds ! "

" There you go, soaring again ! " I cried. " Let us keep on practical subjects. What have the foundry people who built this thing, and the railroad people who brought it down here, thought about its probable use? Have they not guessed something?"

"You may trust the popular mind not to guess flying unless it sees wings! They have imagined this is a new sort of torpedo, sent down here for a private trial in the lake. In fact, the conductor of the freight train, who switched the car off here, asked me in a confidential way if he should get teams and men and help me to launch her? I have fostered this idea, and really had the projectile sent here to carry out that impression."

A more fitting place for an unobserved start could not have been selected, however. All this part of the country is a sandy waste, with a sparse growth of scrub oaks and but little vegetation. There are no farms, and the nearest houses are at Whiting. No one could see our work, except, possibly, the passengers from occasional trains, which rushed by without stopping, and were infrequent at this time of day.

As we were arriving, I stood off at some distance to observe the black object on the open car. It was five feet through, and twenty feet long, not counting the rudder, which was now entirely drawn into the rear end.

"Look's exactly like a cigar," I said. "Sharp and pointed in front, slightly swelled in the middle, and cut squarely off behind. Only it is too thick for its length, of course."

FINAL PREPARATIONS

But the doctor already had the rear port-hole open. This was two feet in diameter, and permitted a rather awkward entrance to the rear compartment. The interior was crowded with boxes, as yet unpacked, containing scientific instruments, tinned foods, biscuits, meat extracts, condensed milk and coffee, bottled fruits, vegetables, and the like. Over these the doctor worked his way to the forward compartment, while I followed him, anxious to explore the interior.

"I will unpack all these goods and put them in their places to-morrow forenoon," explained the doctor. "Here, in my compartment on the left, I have my gravity apparatus, battery cells and the like, and a small table for writing and other work. On the right is the bunk on which I sleep, and under it is the big telescope, neatly fitted and swinging up easily into place before the mica window."

"Has the compressed air been put in yet?" I inquired.

"Oh, yes, that had to be done in the city, where they have powerful air compressors. I would have preferred this purer air out here, but it was impossible. The air we put in only increased the weight of the projectile eighteen pounds, but it will be sufficient for two of us for six months. We were obliged to make the most careful and thorough tests for leaks in the air-chambers; for if there were any of these, our life would leak out with the air."

"And such airless satellites as the Moon will make the most desperate efforts to steal your atmosphere, too!" I added.

"Yes, but we will give them only our foul air as a small stock-in-trade with which they may begin business. But I see my batteries are commencing to work nicely. I think I can lift her now. You go outside and make a hitch with that rope you saw just forward of the middle of the projectile. Then, when I have neutralized her weight, you tow her over beyond that clump of trees you saw near the shore. That will be out of the view of trains."

"Must I concentrate my mind or keep my thoughts fixed on anything?" I asked quizzically.

"Rubbish! Concentrate it on this. If the projectile starts up, don't try to hold her with your little rope. Let go quickly, or you may get uncomfortable holding on!"

I went outside, untied the coil of rope and threw one end over. Meantime the doctor had opened the forward window, so that he might give directions, and I said to him,—

"I can't get the rope under her; she is lying flat on the car."

"Wait a moment and I will lift her for you," he replied. The railroad ties rose a little out of the sand, and there was a slight creaking of the woodwork of the car as the weight came off. Presently the forward end of the projectile rose slowly an inch, two inches!

"That's enough!" I cried, thrusting the rope under, and she settled back gently. Having made my knot, I went out to the other end of the rope, about thirty feet distant. Forgetting the doctor's injunction about not hanging on, I wrapped the rope around my body, worked my feet firmly into the sand, and finally cried out, "All ready!"

There was a faint creaking of the car again, and soon the doctor said, "Pull away!" I threw all my force into the effort and gave a tremendous heave, and tumbled over backwards. Had I not done so, the projectile must have hit me as it glided rapidly from the car, sinking very slowly to the sand about fifty feet away. I scrambled to my feet, went in front again, and easily dragged it along on the sand to an open place just beyond the trees. There the doctor allowed it to settle. It sank into the loose sand about eight inches, remaining steady in this position.

"She works beautifully!" I cried. "How I would like to see her turned loose for a real flight!"

"That will come to-morrow night," said the doctor, crawling out of the port-hole. "But if you will help me remove these boxes from the experimental model, you shall see it lost in the sky." We uncovered and dragged out a small steel thing, about the same shape as the projectile, but less than a foot thick and four feet long. It had a lid opening into its batteries from the top. The doctor entered his compartment to secure some chemicals.

"If you have no further use for this model," I suggested, "why not create a very strong current and let it sail off into indefinite space?"

"Very well; I don't wish to leave it behind me for some one to discover, and I can't take it along. We will send it off for a long trip, and if it falls back it will be into the lake."

"Wait a moment, then! Let's put a good-bye message in it;" and so saying I took an old envelope from my pocket and wrote on the back of it with a pencil in a bold hand: "Farewell to Earth for ever!" Laughing, I put this inside and closed the lid.

Then the doctor turned down a thumb-screw upon a little wire which connected the poles, and stepped back quickly. Presently the forward end began to rise slowly, until it stood upright, but there it hesitated. The doctor stepped forward and gave the thumb-screw a hard turn down, and the model lifted immediately, rising at first gradually, but soon shooting off with the whizz of a rocket over the lake. We watched it as long as we could distinguish its dark outline.

"It will go a long way," said the doctor. "I have never seen it make so good a start. It will lose itself in the lake far from here."

We fastened up the front window and the porthole, and started back to Whiting, where the doctor was to remain all night, so as to begin work early in the morning. Presently, as we walked along, the doctor said,—

"Well, Isidor, now you have seen a practical demonstration of the elementary working of the projectile. You also have some idea of all there is to be discovered up yonder in the red planet. You are the most interested in making and profiting by those discoveries. I want you to consent to go along."

"Haven't you secured a companion, then?" I inquired.

"Yes, I have a friend, a countryman of mine here, who will go wherever I say. He appreciates neither the risks nor the opportunities of the trip, still he will take my word for everything. Yet if I ask him to go I take the responsibility of his life as well as my own. He is not a suitable man, however, and I have really relied on you to come," he insisted.

"My dear doctor, I have every faith in you and in the projectile, and I prophesy a most successful trip. I should like nothing better than the adventure; but you must not count on me; I could not leave my business. There's a fever in my blood that thirsts for it!"

"Your business, indeed! You will never really amount to much till you have left it. It's half a throw of dice and the other half a struggle of cut-throats!"

E

"That is what people say who know nothing at all about it," I retorted. "It occupies a large and important place in the world's commerce. Besides, I could not well leave Ruth and my uncle."

"Isn't it time you did something to make her proud of you, and to be worthy the education which he gave you? You have a chance now to be great. Isn't that worth ten chances to be rich? What would you have thought of Galileo if he hadn't had time to use the telescope after inventing it, but had devoted his time and talent to the maccaroni market? You are one man in ten million; you have an opportunity Columbus would have been proud of! Will you neglect it for mere gold-grubbing? Leave that to the rest of your race and to this money-mad Chicago. You come along with me. Let's make this work-a-day world of ours take time to stop and shake hands with her heavenly neighbours!"

"You tempt me to do it, Doctor! Can you wait two or three days for me?"

"I can, but Mars won't," he answered laconically. "Besides, you must not tell any one that you are going."

"If there are any two things I love, it's a secret and a hurry! I will be here to-morrow night," I exclaimed.

CHAPTER VI

Farewell to Earth

THE next day I quietly bought in my wheat, and told Flynn I was thinking of taking a little vacation. I said I was worn out fighting the contrary market, and told him to run the office as if it were his own until I returned. At home I said nothing about the vacation, for I didn't care to have my stories agree very perfectly. I simply packed a few necessities for the trip in a dress-suit case. My uncle was used to seeing me carry my evening clothes to the Club in this manner, and I casually told him I should remain the night this time.

I could not leave without kissing cousin Ruth good-bye, but this excited no suspicion, as it was a thing I did on every pretext. Then I slipped out and took back streets till I was several blocks away from the house. Taking a closed carriage here, I was driven to the same station and took the same train for Whiting as on the previous evening. I found the doctor awaiting me with a lantern. As we walked down the tracks in the twilight I said to him,—

"I never made so quick a preparation, nor attempted so long a trip. I have left my friends a lot of guessing! Now, how soon shall we be off?"

"Within an hour," he answered. "Mars will not be directly overhead until midnight, but there is a little side trip I wish to make first, to test the projectile before we get too far above the Earth's surface."

The sky was densely cloudy, there was no Moon, and it was already growing very dark. As we began to have difficulty in finding the way, the doctor lighted his lantern. Peering up into the darkness, I said to him,—

"There is not a star visible. How are you to find your way in the heavens a night like this?"

"That is all perfectly easy. We shall soon rise far above those clouds, and then the stars will come out. Besides, I shall show you perfect daylight again before midnight."

"I don't see just how, but I will take your word for it, Doctor. I daresay you have thought it all out, and the whole trip will contain no surprises for you."

"I have tried to think it all out and prepare for everything. But I am certain I have forgotten something. I have a feeling amounting to a dreadful presentiment that I have overlooked something important. I wish you would see if you can think of anything I have omitted."

"The only really important thing I have remembered is half a dozen boxes of the best cigars," I replied.

"Leave them right here in Whiting," he said with emphasis. "We are carrying only a limited supply of pure air, and we cannot afford to contaminate it with tobacco smoke. No, sir, you can't smoke on this trip."

"Then I won't go! Imagine not smoking for two whole months! Do you think I have sworn off?"

"No, not yet. But you must. It pollutes the air, which we must keep clean and fresh as long as possible."

"Now, Doctor, you must let me have a good smoke once a day, just before pumping the air out of my compartment."

"No, not even that. It is impossible to pump all the air out, and what is left mixes back with what is in my compartment. Once contaminated with tobacco smoke, we could never get it perfectly pure again."

"Well, may I smoke on Mars, then? I will take them along for that. But, I warn you, I eat like a farm horse when I can't smoke."

"I have provided plenty to eat, but I know I have forgotten something. Mention something now, mention everything you can think of, so that I may see if it is provided for."

"Have you any money?" I asked. "I have

changed some into gold, and have a fairly heavy bag here."

"Oh, yes, I have some gold and silver money, besides a lot of beads, trinkets, and gaudy tinsel things, such as earthly savages have been willing to barter valuable merchandise for."

"So you are going on a trading expedition, are you?" I asked.

"Not exactly. I leave all that to your superior abilities. But we may find these things valuable to give as presents. Many of them are of tin, and if they do not happen to have that useful metal on Mars, they will be of rare value there."

We had now reached the little grove where the projectile was hidden. I proceeded to open the rear port-hole, saying,—

"Let me look inside, and when I see what you have, some other necessary thing may suggest itself."

"Let me go in first, for I am afraid you will allow the menagerie to escape," he said, as he peered in by the light of the lantern. A diminutive fox terrier barked from the inside, and wagged his tail faster than a watch ticks, so glad he was to see us. The bright light also awakened a small white rabbit that had been asleep in the doctor's compartment.

"You are taking these along for companions, I suppose?"

"Yes, for that and for experiments. We may

reach places where it will be necessary to determine whether living, breathing things can exist before we try it ourselves. Then we shall put one of these out and observe the effects."

"You may experiment on the rabbit all you please, but this little puppy and I are going to be fast friends, and we shall die together; shan't we, Two-spot?"

"Why do you call him Two-spot? There is only one spot on him, and his name is *Himmelshundchen*."

"Rubbish! The idea of such a long, heavy name for such a little puppy! I shall call him Two-spot because he is the smallest thing in the pack. Heavenly-puppy, indeed!"

The doctor had entered and lighted a small gas jet, supplied on the Pintsch system from compressed gas stored in one of the chambers. The rear compartment, which was to be mine, looked half an arsenal and half a pantry. On the right side a cupboard was filled with newly-cooked meats. I remember how plentiful the store looked at the time, but, alas! how soon it vanished and we were reduced to tinned and bottled foods! There was a cold joint of beef, a quarter of roast mutton, three boiled hams and four roast chickens.

On the left, folding up into the concavity of the wall, like the upper berth of a Pullman sleeping car, was my bunk. On the walls not thus occupied the arms were hung. There were two repeating rifles, each carrying seventeen cartridges; two large calibre hammerless revolvers; two long and heavy swords, designed for cleaving rather than for stabbing; two chain shirts, to be worn under the clothing to protect against arrows; and finally two large shields, made of overlapping steel plates and almost four feet high. The doctor explained to me that the idea was to rest the lower edge of these on the ground and crouch behind them. They were rather heavy and cumbersome to be carried far, and were grooved in three sections, so that they slipped together into an arc one-third of their circumference.

I examined everything closely and asked a hundred questions, but the doctor seemed to have provided for every necessity or contingency.

"Let us waste no more time," said I. "If we have forgotten anything, we must get along without it. All aboard! What is our first stop?"

"The planet Mars, only thirty-six million miles away, if we are successful in meeting him just as he comes into opposition on the third day of August. This is the most favourable opposition in which to meet him for the past quarter of a century. Back in the year 1877 he was only about thirty-five million miles away, and it was then that we learned most that we know of his physical features. But we shall not have a more favourable time than this for the next seventeen years."

"Still it seems like nonsense to talk about travelling such an incomprehensible distance, doesn't it?" I ventured.

"Not at all!" he replied positively. "If the Earth travels a million miles per day in her orbit, without any motion being apparent to her inhabitants, why should we not travel just as fast and just as unconsciously? We are driven by the same force. The same engine of the Creator's which drives all the universe, drives us. When we have left the atmosphere we shall rush through the void of space without knowing whether we are travelling at a thousand miles per minute or standing perfectly still. Our senses will have nothing to lay hold on to form a judgment of our rate of speed. But if we make an average of only five hundred miles per minute we shall accomplish the distance in about fifty days, and arrive soon after opposition."

"But have you given up stopping on the Moon?" I asked. "I had great hopes of making those rich discoveries there."

"We must leave all that until our return trip. I have chosen this starting time in the dark of the Moon in order to have the satellite on the other side of the Earth and out of the way. She would only impede our progress, as we wish to acquire a tremendous velocity just as soon as we leave the atmosphere. We must accelerate our speed as long as gravity will do it for us. When we can no longer gain speed, we shall at least continue to maintain our rapid pace.

"But if we stopped on the Moon, we should only have her weak gravity to repel us towards Mars, and we could make but little speed. On our return, the stop on the Moon will be a natural and easy one. We shall be near home and can afford to loiter."

While the doctor was saying this, he had been busy making tests of his apparatus. He now called me to see his buoyancy gauge, which was a half-spherical mass of steel weighing just ten pounds. It was pierced with a hole at right angles to its plane surface and strung upon a vertical copper wire. Small leaden weights, weighing from an ounce to four pounds each, were provided to be placed upon the plane surface of the steel. The doctor explained its action to me thus:—

"The polarizing action of the gravity apparatus affects only steel and iron, and has no effect upon lead. Therefore, when the current is conducted through the copper wire into the soft steel ball, it will immediately rise up the wire, by the repulsion of negative gravity. Now, if the leaden weights are piled upon the steel ball one by one, until it is just balanced half way up the wire, our buoyancy is thus measured or weighed. For instance, with the first two batteries turned in we have a buoyancy a little exceeding one pound. That means, we should rise with one-tenth the velocity that we should fall. Turning in two more batteries, you see the buoyancy is three pounds, or our flying speed will be three-tenths of our falling speed. With all the batteries acting upon the gauge, you see it will carry up more than ten pounds of lead, because the pressure of the air is against weight and in favour of buoyancy. So long as we are in atmospheres, then, it is possible to fall up more rapidly than to fall down; but, on account of friction and the resultant heat, it is not safe to do so."

"So we have been doing the hard thing, by falling all our lives, when flying would really have been easier!" I put in.

"We have been overlooking a very simple thing for a long time, just as our forefathers overlooked the usefulness of steam, being perfectly well acquainted with its expansive qualities. But let us be off. Close your port-hole, and screw it in tightly and permanently for the trip. Then let down your bunk and prepare for a night of awkward, cramped positions. We shall be more uncomfortable to-night than any other of the trip. You see, when we start, this thing will stand up on its rear end, and that end will continue to be the bottom until we begin to fall into Mars. Then the forward end will be the bottom. But after the first night our weight will have so diminished that we can sleep almost as well standing on our heads as any other way. Within fifteen hours you will have lost all idea which end of you should be right side up, and we will be quite as likely to float in the middle of the projectile as to rest upon anything."

My bed was hinged in the middle, and one end lifted up until it looked like a letter L, with the shorter part extending across the projectile and the longer part reaching up the side. I could sit in it in a half reclining posture. The doctor then pulled out a fan-like, extending lattice-work of steel slats, to form a sort of false floor over the port-hole. This was full of diamond-shaped openings between the slats, so that the view out of the rear window was not obstructed. Then he did the same to form a false floor for his compartment. Finally he said to me,—

"Now, if you are all ready, I will stand her on end;" and by applying the currents to the forward end only he caused her to rise slowly until she stood upright. The cupboard in my compartment and the desk in his end were each hung upon a central bolt, and they righted themselves as the projectile stood up, so that nothing in them was disarranged. I was sitting on the lower hinge of my bed, clutching tightly and watching everything, when the doctor called to me to turn the little wheel which operated a screw and served to push out the rudder.

"But the whole weight of the projectile is now on the rudder," I objected.

"You will have to make over all your ideas of weight," he said, with some impatience. "Run the rudder out. The gauge shows an ounce of buoyancy, which is nearly enough to counteract all the dead weight we have. You can lift the rest with the rudder-screw."

And, true enough, it was perfectly easy to whirl the little wheel around which made the rudder creep out. There was a steering wheel in the doctor's compartment and one in my own. He set it exactly amidships, and told me to prepare for the ascent. I turned out the gas in my compartment and crouched nervously over the port-hole window to watch the panorama of Earth fade away.

"Here go two batteries!" he cried. I held on frantically, expecting that we would leap into the heavens in one grand bound, as I had seen the model do. But we began to rise very slowly, a foot and a half the first second, three feet the next, and so on, as the doctor told me afterwards. It was all so slow and quiet that I was suddenly possessed with a fear that after all the projectile was a failure. Had a balloon started so slowly, it would never have risen far. This fear held me for only a minute, for when I looked down again, the landscape below was beginning to look like a dim map or a picture, instead of the reality. The doctor was steering to the northward, directly over the lake. I could see its great purple, restful surface below me, but more plainly could I discern the outline where its silvery edge bathed the white sands of the shore. Following this outline I could see a web of railroads, like ropes bent around the lower end of the lake. The night was too dark to see it long. The hundreds of huge oil tanks of Whiting had now disappeared, and I could see only the flaming tops of the iron furnaces of South Chicago. Suddenly they went out in an instant, as if a thick fog had smothered them, and there was a long minute of pale mist; and then suddenly a bright blue sky, the twinkling stars and a veil of grey shutting off all view of the Earth.

"We have passed through the clouds," said the doctor cheerily. "What does the barometer register?"

I looked, and was astonished to see the mercury down to fifteen. I asked him if he thought the barometer might be broken.

"No, that is quite right," he replied. "That is half the surface pressure, which shows that we are two and a half miles high. I have four batteries in, and we are going at a constantly increasing speed now."

I could easily believe it, for the wind howled around my compartment and whistled over the rudder aperture in a most dismal way. Whenever the rudder was changed, there was a new sound to the moaning. Still, as I looked back at the clouds, I saw that no wind was moving them. It was not wind, but only the air whistling as we rushed through it.

"Watch the barometer, and let me know the

exact time when it registers seven and a half inches," said the doctor. "We shall be five miles high then, and we started at nine o'clock to a second."

I noted the rapidly sinking mercury and opened my watch. When it was just at seven and a half, I looked at the watch, and it said half a minute after nine. Knowing that could not be correct, I held it to my ear and discovered it was stopped. I attempted to wind it, but found it almost wound up.

"Something wrong with my watch, Doctor. You will have to look."

"Half a minute after nine, that can't be right!" he exclaimed. Then as the truth flashed upon him he added,—

"There is the first thing I have overlooked! Our watch springs are steel, and the magnetic currents affect them. It is strange I did not think of that, for I knew a mariner's compass would be of no use to us in steering on account of the currents. For that reason I have risen above the clouds so as to steer by the stars. I am making for the North Star yonder, now."

"We will have to get back to the same primitive methods of measuring time," I put in. "Neither weight clocks nor spring clocks would have been of any account. And an hour glass would tell a different tale just as gravity varied. We will have to rely on the Moon and stars, and it may be rather awkward." But I did not then appreciate how awkward it would be when even the markings of day and night would be taken away from us.

"We can count our pulse or go by our stomachs," said the doctor, who was really disappointed at having forgotten anything. But he was destined to get used to that. Presently he inquired,—

"What is the barometer now? Perhaps we are high enough for the present."

"There is scarcely two inches of mercury in the tube!" I cried out.

He hesitated for a moment as if calculating, and then said,—

"That makes us ten miles high. Work the rudder gradually very much farther out for this thinner atmosphere, and we will try falling awhile, with a long slant to northward."

And so saying, the doctor detached all the polarizing batteries, and I could hear the monotonous howling of the wind die down; and the whistling ceased altogether as the feeble resistance of the rarefied air slowly but surely overcame our momentum. As we began to fall, the doctor turned the rudder hard down, in order to give us a long sailing slant. This modified the position of the projectile so that it lay almost flat again, with a dip of the forward end downward.

"Lie down and have a nap while she is in this comfortable position," he said to me. "When you waken, I shall have a surprise for you."

CHAPTER VII

The Terrors of Light

I WAS weary from the trials of the day on Earth, and fell asleep easily. It was the red sunlight streaming in at the port-hole that awakened me. I thought I had slept but a very short time, but the night was evidently over. As soon as the doctor heard me moving, he cried out to me,—

"Here is the daylight I promised you. Did you ever see it at midnight before?"

"How do you know it is midnight? It looks more like a red sunset to me," I said, for the sun was just in the horizon.

"The sun has just set, and is now rising. It did not go out of sight, but gradually turned about and began to mount again. That is how I know it is midnight."

"Sunset presses so closely upon sunrise that night is crowded out altogether. Then this must be the land of the midnight sun that I have read about?"

"Yes, we are very near the Earth again, and this is far inside the arctic polar circle, where the sun never goes down during summer, but sets for a long night in the winter. I have kept far to the westward to avoid the magnetic pole, which might play havoc with my apparatus."

"Then your little side-trip is ——"

"To the North Pole, of course!" he cried triumphantly.

How simple this vexed problem had become, after all! It had worsted the most daring travellers of all countries for centuries. Thousands upon thousands spent in sending expeditions to find the Pole had only called for other thousands to fit out relief expeditions. Ship after ship had been crushed, life after life had been clutched in its icy hand! But now it had become an after-thought, a side-trip, a little excursion to be made while waiting for midnight! And it is often that such a simple solution of the most baffling difficulties is found at last.

The doctor had been observing his quadrant, and was now busy making calculations. He called me up to his compartment.

"Longitude, 144 degrees and 45 minutes west; Latitude, 89 degrees 59 minutes and 30 seconds north. That is the way it figures out. We were half a mile from the Pole when I took my observation. We must have just crossed over it since then."

"Go down a little nearer, so we may see what it looks like!" I said excitedly.

"I dare not go too close to all that ice, or we may

freeze the mercury in our thermometer and barometer. We must keep well in the sunlight, but I will lower a little."

What mountains of crusted snow! What crags and peaks of solid ice! It was impossible to tell whether it was land or sea underneath. Judging by the general level it must have been a sea, but no water was visible in any direction. The great floes of ice were piled high upon each other. A million sharp, glittering edges formed ramparts in every direction to keep off the invader by land. How impotent and powerless man would be to scale these jagged walls or climb these towering mountains! How absolutely impossible to reach by land, how simple and easy to reach through the air! The North Pole and Aerial Navigation had been cousin problems that baffled man for so long, and their solution had come together.

"Empty a biscuit tin to contain this record, and we will toss it out upon this world of ice, so that if any adventurer ever gets this far north he may find that we have already been here," said the doctor, bringing down a freshly-written page for me to sign. It read as follows:—

"Aboard Anderwelt's Gravity Projectile, 12.25 a.m., June 12th, 1892. The undersigned, having left the vicinity of Chicago at nine o'clock on the evening of June 11th, took bearings here, showing that they passed over the North Pole soon after midnight. Then they took up their course to the planet Mars.

"(Signed) Hermann Anderwelt.
"Isidor Werner."

This was duly enclosed in the biscuit tin, which I bent and crimped a little around the top so that the cover would stay on tightly. Then I learned how such things were conveyed outside the projectile. A cylindrical, hollow plunger fitting tightly into the rear wall was pulled as far into the projectile as it would come. A closely fitting lid on the top of the cylinder was lifted, and the tin deposited within. The lid was then fitted down again, and the plunger was pushed out and turned over until the weight of the lid caused it to fall open and the contents to drop out. The tin sailed down, struck a tall crag, bounded off, and fell upon a comparatively level plateau. The cylinder was then turned farther over, causing the lid to close, and the plunger was pulled in again. I remember how crisply cold was that one cubic foot of air that came back with the cylinder. My teeth had been chattering ever since I wakened, and I had been too excited to put on a heavier coat.

"What is the thermometer?" asked the doctor. It was a Fahrenheit instrument we were carrying.

"Thirty-eight degrees below zero, and still falling!" I told him.

THE TERRORS OF LIGHT

"Then we must be off at once, and at a good speed, to warm up. Now say a long good-bye to Earth, for it may be nothing more than a pale star to us hereafter."

The doctor steered to westward as he rose steadily to a height of about ten miles. Then he fell with a long slant to the south-west. He was working back into the darkness of night again. We had lost the sun long before we started to rise again.

"We are now well above the Pacific Ocean, about fifteen hundred miles north-west of San Francisco," said the doctor, consulting his large globe.

"It seems to me you cross continents with remarkable ease and swiftness. From Chicago to San Francisco alone is almost three thousand miles," I ventured.

"But we have been gone four hours, and if we had simply stood still above the Earth for four hours it would have travelled under us about four thousand miles, so that San Francisco would already have passed the place where we started."

"Then one only needs to get off somewhere and remain still in order to make a trip around the World!" I exclaimed.

"You are quite right, and travelling upon the Earth's surface is the most awkward method, because it is impossible to take advantage of the Earth's own rapid motion. Around the World in eighty days was once considered a remarkable feat, but if we were to travel steadily westward we should make the circuit in very much less than twenty-four hours. The motion of the Earth upon its axis is such an immense advantage that if we were only going from Chicago to London, the trip could be more easily and quickly made by going to the westward some twenty-one thousand miles, rather than going directly eastward less than four thousand miles. For going eastward we should have to travel a thousand miles an hour in order to keep up with the Earth. It is questionable whether we could make that speed tacking up and slanting down."

"Then we shall have to follow the course of Empire, always westward!" I laughed.

While we were talking thus, the whizzing and whistling of the wind, which had been at first very loud and hissing, had gradually died down. I looked at the barometer, and reported that there was scarcely three-eighths of an inch of mercury in the tube.

"We are practically above the atmosphere, then," said the doctor, turning in all the batteries. He tried the rudder in the ether, and found it turned her when fully extended and turned rather hard over.

"I tried to sleep this morning at Whiting to prepare for to-night's work," said the doctor presently; "but I find I am getting uncontrollably drowsy. Come up, and I will show you the course we must keep, and then I will lie down to get a little rest."

I mounted to his compartment and gazed through the telescope at Mars, looking like a little, red baby-moon, floating in one side of the blue circle.

"Keep him always in view, but in the edge of the field like that," said the doctor. "We must always steer a little to the right of him—that is, a little behind him."

"But he travels around the sun in the same direction the Earth does," I objected. "I should think we ought to aim a little ahead of him, or to the left, to allow for his motion forward in his orbit."

"That looks reasonable at first sight, doesn't it?" said the doctor. "But a little learning is a dangerous thing. I will explain to you why we must steer a little behind him after I have had my nap. I am too sleepy now;" and he finished with a yawn.

He soon fell asleep, and I was left alone to think over the events of the day and the still more strange happenings of the night. It hurt my eyes to look long through the telescope, so I closed them and gave free rein to my thoughts.

How soon will it be morning? How shall I know when it *is* morning? That term "morning" applies only to the surface of revolving planets. I had just seen the morning come at midnight, and then the darkness of night fall again directly after morning. After all, what are night and morning? The one is a passing into the shadow of the Earth, and the other is simply the emerging into the light. They depend on a rotation, and we shall know no more of them until we land on a revolving planet again. But which shall we have on the trip, night or daylight? Naturally we would very soon emerge from the little shadow cast by the Earth. It had taken us but an hour or two to travel out of it into the daylight and then back into the darkness again. Even if we did not leave it, the Earth would move on and leave us.

And what then? Nothing but uninterrupted, untempered, unhindered daylight! Eternal, dazzling, direct sunlight, unrelieved by any night, unstrained through any clouds! This deep blue of the starry night would be succeeded by the hot, white light of a scorching, gleaming Sun. And then (the thought chilled my bones as it fell upon me!), then how would we see Mars? How would we see any star, or perchance the Moon? Even the Earth might be drowned in that sea of everlasting, all-engulfing brilliancy! Nothing in all the Universe would be visible but the beaming Sun, and he too blindingly bright to look upon.

As the truth of all this took hold of me, it filled me with a growing terror. At any moment we might emerge from this grateful shadow of the Earth, and then we would be lost, drowned, engulfed in a blinding, sight-suffocating light! In desperate terror I looked around toward the doctor, as if for assistance. He was sleeping peacefully. He had never thought of it! *This* was the great thing he had overlooked! Even at starting he had a dreadful presentiment of it.

He was a great man, and his discovery a wonderful one; but here was the trouble with it. He had solved the question of navigating space, but the sunlight! the dazzling, burning, terrible sunlight! how was he to navigate that? It was simply impossible! We would have to turn back before we emerged into it. We would have to retrace our path while we were still in the grateful shadow. Ah, the blessedness of night after all!

Then slowly and cautiously, so that I might not waken him, I crept down to the rear window to see how far away the Earth was. We were at so great a distance that I could see the whole outline of it, as a great dull globe filling all the view behind us. And as I looked again I started and uttered a cry! A thin sickle of bright, white light glimmered over the whole eastern edge of it, like the first glimpse of the new Moon, but a hundred times larger! It was the sunlight! It must be creeping around the eastern edge, and would soon engulf us.

The doctor had been aroused by my cry. Not seeing me in his compartment, he had gone at once to the telescope.

"What is the matter?" he said. "You have lost the course a little." And as I peered out of my port-hole I saw that narrow sickle of light grow thinner and thinner, and finally go out. Had I imagined it all? No, I had seen it.

"Ah, Doctor, I am so glad you have wakened. I am frightened, terrified, by the light!"

CHAPTER VIII

The Valley of the Shadow

"LIGHT! Where have you seen any light?"

"I saw the Earth begin to shine like a New Moon on the eastern edge, but——"

"Ah, that *was* a danger signal. I am glad you awakened me. But you are actually pale and trembling! There is no danger if you keep the course. You see, that rim of light has faded and disappeared since I corrected the course."

"Yes, but you cannot keep in this little Earthly shadow much longer; and what can we possibly do when we emerge into the fathomless, trackless effulgence of eternal sunshine? Let us turn back before we plunge into it," I pleaded.

"So that is what terrified you! Well, you have hit upon one of the greatest difficulties of the trip; but it is far from insurmountable. We will not turn back yet, especially as we have started in the most opportune time. You have mentioned this 'little shadow.' It is eight thousand miles wide at the surface of the Earth, and gradually, very gradually, tapers down to nothing far out in space. Have you ever calculated how far it reaches?"

"No," I answered. "But we moved out of it and back into it at the surface very easily, and besides, as the Earth moves forward in its orbit, the shadow will leave us."

"This little shadow is eight hundred and fifty-six thousand miles long, and we will never leave it as long as it lasts!" exclaimed the doctor. "Just at this time it points like a long arrow out in the direction of Mars. It is moving gradually as the Earth moves and hourly correcting its aim. At opposition time it will point directly and unerringly at Mars. Therefore it is a way prepared, surveyed, and marked for us through the all-enveloping sunlight, which otherwise would be dreadful enough."

"But how can we be sure of keeping in it? It is rapidly narrowing as it reaches farther out."

"I see I should have explained that to you before I went to sleep, and saved you this fright. The shadow now points behind Mars, as it is many days yet before it overtakes that planet in opposition. That is why I told you to steer always a little behind the planet. But you went a little out of the course, and immediately something warned us. That rim of light on the east of the Earth was notice to us that we were not in the centre of the shadow, but bearing too far to the left. We must keep absolutely in the dark of the Earth, with no light visible on either side of it. If a thin rim should appear on one side, we must turn toward the other until it is all dark again."

"Grant that this shadow is so enormously long, yet it is only scarcely one-fortieth of the distance to Mars," I objected. "After we emerge from it, what then?"

"With the aid of my telescope we shall probably be able to see the Earth as an orb, half or quarter as large as the Moon usually appears to us, and to observe its phases until we are several million miles from it. We must continue to keep the rim of light, which will then surround it, equal on all sides."

"Ah, but I am afraid," I interrupted, "that as soon as we pass out of this shadow the sunlight will be so bright that we cannot see any planets, not even the Earth. You know we cannot see the Moon only a quarter of a million miles away when the sun shines."

"In that case we must move the telescope to your window, put on a darkened lens, and steer so as to keep the Earth as a spot in the middle of the Sun. It must appear to us as Venus does to the Earth when she is making a transit across the face of the sun. But by our continual shifting we prevent the Earth from making a transit, and hold it as a steady spot in the centre of the Sun. This we can do for many, many million miles, continuing until we have reached the vicinity of Mars.

"And you must also remember," continued the doctor, "that the brighter the light the darker will be the shadow. Now, this projectile is a perfectly black, non-reflecting object five feet wide. It will cast a shadow in front of it five hundred feet long. When we are comparatively near Mars my telescope, situated in the miniature night cast by the projectile, will find the planet, and we can then steer directly for him. If we should chance within eighty thousand miles of him, he would attract us to him in a straight line. But we shall not rely upon chance. Moreover, when we are as near to him as that, the light and heat of the Sun's rays will have decreased sixty or seventy per cent. When Mars is farthest from the Sun, he receives only one-third as much light as the Earth does. But he is now almost at his nearest point to the Sun, and receives half as much light."

"Well, you certainly have a pretty clear idea of how to steer the course all the way, Doctor. And I was hasty enough to think you had overlooked this entire phase of the subject!" I ejaculated.

"Indeed, I have thought of it very much. And we should not enjoy all these advantages if we had not started just before opposition. At any other time the Earth's shadow would not point toward Mars, nor would the transit of the Earth over the Sun be of any use to us."

"All this reassures me greatly," I replied; "but I shall keep a close watch from my rear window for danger lights on the Earth."

"It must be time for breakfast," put in the doctor. "Will you see how tempting a meal you can prepare?"

There was one reservoir built inside the compartments, from which we drew cool water, and another built next to the outer steel framework, from which we could draw boiling water. As this tank was connected with the discharge pipe of the air-pump, and thus with the exterior, I was disgusted to find that, although the water boiled furiously, and was rapidly wasting away in steam, it did not become hot enough to make good beef tea. The heat escaped with the steam at a comparatively low temperature, so that I was compelled to boil water over my gas jet for the meat extract, which we drank instead of coffee. I also prepared some sandwiches of roast beef and cold ham, and with great relish we began our diet of ready cooked foods, which was to continue for so long.

After this meal I felt quite sleepy, for I had enjoyed but three hours' rest. The doctor saw my yawns and told me to turn out the gas and have a long doze, and I was glad enough to do so.

I must have slept soundly for an hour or two, and then I remember dozing and rolling lazily in my bed, as I usually did at home on Sunday mornings. During my previous nap the bunk had seemed hard and cramped, and I had privately grumbled at the doctor for overlooking personal comforts; but now I felt that luxurious sensation of sleeping on soft mattresses and yielding springs, though of course I had neither. I do not know how soon I should have thoroughly awakened had I not lifted my hand to rub my eye, and unwittingly dealt myself a stinging blow in the face. This roused me.

But what was the matter with that arm? It was as it had once been in a nightmare, when it felt detached from its place, and moved lightly and without effort, like a bough in the wind. I pinched it with my other hand, and it was quite sensible to the pain. In fact, the other arm was now acting in the same queer way. I arose in bed quickly to see what was the matter, and the upper part of my body bent violently over and struck against my knees. Then my effort to take an upright position threw me on my back again. Evidently my muscles were not working as they were when I went to bed. They must be overexcited and over-active. I immediately thought of my heart as the principal and controlling muscle, and in my eagerness to feel its beating my hand dealt me a slap in the chest. These blows, though rapid, did not seem to hurt as much as they ought, after the first stinging sensation. I found my heart was beating regularly enough.

"Doctor!" I cried out presently, more to test

my voice than for anything else. It sounded perfectly natural, and my vocal chords were not over-stimulated or abnormal.

He came half way down from his compartment soon after hearing me, and rested his elbow against one side of the aperture between the compartments, leaning against the other side easily. He had a scale made of heavy coiled spring in his hand.

"I wish to calculate our distance from the Earth," he said. "Do you mind weighing yourself on these scales?" and he held the spiral down toward me.

"You can't support my weight!" I exclaimed, and springing up from the bed I bumped my head against the partition between the compartments, eight feet above my floor. I grasped the lower ring of the scale he held down and lifted up my feet. It seemed as if something were still supporting me from below, for scarcely one-tenth my weight had fallen upon my hands.

"You weigh twenty and a half pounds," he said, and then inquired, "What did you weigh on Earth?"

"One hundred and eighty-five pounds," I answered, just beginning to understand that our greatly increased distance from the Earth had much reduced her attraction for us.

"That is disappointing," he answered, "for we are only eight thousand miles from home; but our velocity is still constantly increasing."

G

"I would like to buy things here and sell them at the surface," I exclaimed.

"You wouldn't make anything by it if you used the ordinary balance scales," replied the doctor.

Try as hard as I would, I could not accustom my muscles to these new conditions. They were too gross and clumsy for the fine and delicate efforts which were now necessary. I was constantly hitting and slapping myself, though these blows scarcely hurt, and never resulted in bruises. I attempted a thorough re-training of my muscles, which was to all intents an utter failure, for weight continued diminishing much more rapidly than my stubborn muscles could appreciate. After another eight thousand miles, which were quickly made, we had but one twenty-fifth our usual weight, which reduced me to seven pounds. And for most of the trip we weighed practically nothing, suffering many inconveniences on that account.

CHAPTER IX

Tricks of Refraction

THE doctor figured out that we should be quite insensible to any weight when we were seventy-five thousand miles from the Earth. At fifty thousand miles I would still weigh a pound, and when we had finished the first million miles, the entire projectile, with its two occupants and all its dead weight, would weigh considerably less than an ounce. That was a mere start on the enormous trip ahead of us; but when that distance was reached, we could no longer count upon terrestrial gravity for accelerating our speed. We must travel with our accumulated momentum, unless by that time the Sun should have taken the place of the Earth, and with his vaster forces continue to repel us Marsward.

As we sat talking the doctor grew weary, and soon unconsciously dropped asleep. I left him to enjoy his rest, and, tossing a scrap of ham bone to Two-spot, I went up to take my place at the telescope.

Mars seemed to be exactly in the right part of the field. I surveyed the starry stretches ahead with a feeling a little akin to fear. I was queerly affected by the vast expanse of loneliness outside, and by the deathly quiet prevailing both without and within. There was not the slightest whizzing or whistling now. We might be hanging perfectly motionless in space for all I knew. The batteries made no sound either. I could hear only the low, regular breathing of the doctor as he slept, and the slight crunching of Two-spot on his bone. Presently I thought of looking for the danger lights, but I looked through the telescope instead, and saw the little red planet in his proper place.

What a vast distance we were from any planet! If anything were to happen to us, no one on Earth or in the heavens would ever know of it. I had never been homesick, but a very little would have made me Earthsick just then. I did not like the upper end of the projectile because I could not look back at the home planet. I wondered if it was all dark back that way, or if those warning lights had begun to appear. That idea seemed to haunt me. I touched the steering wheel just a little while I kept my eyes on Mars. He moved slightly in the field at once. Then I turned the wheel back until he took his former place. It was reassuring to know how easily the projectile minded her great rudder, which was now fully extended like an enormous wing. This made me feel that we were masters of the situation, that all this vast space was as nothing to us, that any planet in the heavens must mind us, and that though Earth was driving us away, she must draw us back if we willed it. More than that, she would warn us of all dangers. Perhaps she was sending that warning now. I had promised to look out for it. I felt that I must go down. I crept softly past the doctor and stooped over the port-hole. My eyes had scarcely found the Earth in the darkness when I drew back quickly and clapped my hand over my mouth to prevent a cry escaping me. Then I looked again more closely. There was no small illuminated portion of the surface this time, but a great smear of light just outside the edge of the Earth. It was of a dull red colour, with rainbow tints around the edges, and was much the shape of a great umbrella held just above one quarter of her surface to westward.

I gave the steering wheel in my compartment a sharp turn in the direction which should cause the light to disappear. Then I crouched and looked again, but instead of being reduced in size the light broadened and swelled. It was as if one edge of the umbrella were left against the Earth's surface, and then the umbrella was being turned gradually around until it faced me and formed an enormous disc, apparently a third as big as the Earth. Then, as it slowly moved outward, its edge seemed to cleave to the Earth's, as two drops of water do when about to separate. Finally, it detached itself entirely, and stood as a great muddy red orb a little to the west of and above the Earth. It filled me with dismay to see all this happen after I had turned the rudder in the direction which should have corrected our course. In desperation I gave the wheel an additional hard turn and looked again. At last the great red patch was shrinking; slowly it diminished, and finally disappeared. But just as I was breathing a sigh of relief, I noticed the white sickle of light on the east side that I had seen before; only it was increasing most threateningly now. Yes, it was assuming the same umbrella shape and detaching itself a little from the eastern edge of the Earth. There was still a narrow rim of bright white light on the Earth, and this dimmer umbrella shape was faintly separated from its edge. Its outlines were marked by flashes of rainbow colours, as had been the case on the other side. I sprang to the wheel and gave it several frantic turns back the other way. Then I ran up to the telescope for a hurried view, and Mars was nowhere to be seen! I hastened back to the wheel and gave it a vicious additional turn. I was determined to prevent this umbrella from opening at me! And true enough it ceased enlarging, and gradually shrank and settled back upon the surface of the Earth. Then slowly it faded and disappeared, as it had done before when the doctor had corrected the course. I eased back the wheel and went to look for Mars again, but he was not in the field. As I returned I brushed unconsciously against the doctor in my excitement. He roused himself, sat up, and watched me peering out of the port-hole. I was gazing at a new appearance.

"There it is again!" I cried, for below the Earth and to westward a pale white disc came into view all at once, not gradually, as if emerging from behind the Earth, but springing out complete and detached.

"Doctor!" I said, catching him by the arm and pulling him down to the port-hole, "what is that?"

"That? That is the Moon, my boy. Has it excited you so much?"

"Yes; I have been trying to dodge it. But you had better look to the wheel," I cried.

He ran up to the telescope, and I heard him exclaim, "*Donnerwetter!*" half under his breath. But with a few careful turns of the wheel he found the planet again, and moved him to the right part of the field. Meanwhile the Full Moon shone on us with its pale glimmer. But a thin rim of it next to the Earth gleamed brightly with rich silver light.

"I thought you said we had started in the dark of the Moon. I thought it was behind the Earth," I interposed.

"That is the New Moon just emerging. It will probably not be seen on the Earth until to-morrow night, but as we are at a greater distance we see it first," replied the doctor.

"But that is not a New Moon, it is a Full Moon, which should not be seen for fourteen days yet," I objected.

"Pardon me, it *is* a New Moon," he insisted. "That inner rim of brightness is all the sunlight she reflects. The paler glimmer is Earth-light, which she reflects. When she is really a Full Moon, she will be perfectly dark to us."

Then I explained to him the first umbrella appearance, and its gradual swelling and final disappearance.

"Rainbow colours around the edge and a gradual changing of the shape, you say? That means refraction. The Earth's atmosphere has been playing tricks on you. The umbrella of dull red light was a refracted view of the Moon before she really came into sight. Rays of light from the hidden Moon were bent around to you. Then, as she gradually moved from behind the Earth, her appearance was magnified by the convex lens formed by the atmosphere, bent over that planet. Presently it diminished and went out altogether, you say?"

"Yes, but that was because I steered away from her," I replied.

"No; you could hardly lose her so easily," he

answered. "Did you ever try holding an object behind a water-bottle or a gold-fish jar? There is a place near the edge of the jar where a thing cannot be seen, though the glass and water are perfectly transparent. The rays of light from the object are bent around, through the glass and water, away from the eyes of the observer. It was like that with the Moon when she disappeared. She was really drawing out from the Earth all the time. Finally, when her light passed beyond the atmosphere altogether, she became suddenly visible in a different place and shining with another colour. What we see now is the real Moon in her true place. The other appearances were all tricks of refraction."

"But when I had turned away," I explained, "there came a thin rim of bright light on the other side of the Earth, and a gradually appearing umbrella shape there too."

"Ah, then you steered far enough out of your course to see part of the illuminated surface of the Earth. That was the real danger light. And if it began to assume the umbrella shape, detached from the Earth, that was due to atmospheric refraction of sunlight. This great shadow we are travelling in has an illuminated core, which we shall encounter when we have proceeded a little further. I tell you of it now, so it may not give you another shock. Have you ever noticed the small bright spot which illuminates the centre of the shadow cast by a glass of water? That is partly the same as the core of light which exists in the heart of this shadow. Rays from the sun, passing on all sides of the Earth, are refracted through the atmosphere and bent inward. You must have steered over into some of these rays just now, and then turned back from them. Somewhat farther on all these refracted rays will meet at a common centre, which they will illuminate, and we shall have an oasis of rainbow-tinged sun-light in this great desert of shadow. The sun will then appear to us to be an enormous circle of dull light entirely surrounding the Earth."

"I don't fancy running into that at all," said I. "Can't we avoid it by steering out?"

"Avoid it!" exclaimed the doctor. "We must investigate it, and photograph the peculiar appearance of the sun. Light seems to have more terrors for you than anything else just now. You must get over your rush-and-do tendency; you must stifle your emotions and impulses, and learn to think of things in a more calm and scientific manner."

"But that is not so easy for me, Doctor. Whenever I am left alone, a feeling of dread possesses me. I am used to having many people, bustling noises, and confused movement all about me. The silence of Space stifles me, and the loneliness of the ether oppresses and overcomes me strangely."

"I prescribe a change of air for you," answered the doctor. "You will do better in a rarer atmosphere. Let us send what we have been breathing back to Whiting, and make a new one to suit ourselves."

CHAPTER X

The Twilight of Space

"SHALL I come up into your compartment for the operation?" I asked.

"No; for this first time we will pump out my compartment, as I wish to observe from the rear port-hole the action of the air which we set free."

The bulkhead, with its bevelled edge, was therefore fitted into the opening between the compartments, and I took the first turn at the lever handle of the air-pump, while the doctor observed from the window. I had given the handle less than a dozen vigorous strokes when the doctor suddenly exclaimed,—

"Stop! Wait a moment;" and he began pulling at the bulkhead, which was already rather tightly wedged in by the air pressure. "I have left the rabbit inside," he said, when he found breath to speak. And poor little bunny's heart was beginning to beat fast when he was rescued.

Then we began again. The doctor watched the escaping air for some time, evidently forgetting that I was at all interested in it.

"All quite as I expected," he said at last. "Only I had forgotten about the snow."

"Nothing will ever be very new or interesting to you," I put in; "but pray remember I am here, and rapidly getting empty of breath and full of curiosity."

Then he relieved me at the pump handle, and this is what I saw from the port-hole: The air escaping from the discharge pipe of the air-pump was visible, and looked like dull, grey steam. Immediately on being set free it swelled and expanded greatly, and sank away from us slowly. But at the instant of its expansion the cold thus produced froze the moisture of the air into a fine fleecy snow, which lasted but a second as it sank away from us and melted in the heat, which the thermometer showed to be close upon ninety-five degrees. This miniature snowstorm was seen for an instant only after each down motion of the pump handle.

"Where is this air going?" I inquired. "The little clouds of it seem to drop away from us like lead; but that must be because of our speed."

"It is falling back to the Earth, to join the outer layer of rare atmosphere there. If we had a positive current instead of a negative one, the air would not leave us, but we should gradually be surrounded by an atmosphere of our own, which we should retain until some planet, whose gravitational attraction is vastly stronger than ours, stole it from us. When we begin to fall into Mars, we shall acquire such an enveloping atmosphere; and we can draw upon it and re-compress it if our inner supply should become exhausted."

"If this air is falling home to earth," said I, "we could send messages back in that manner."

"We can drop them back at any time, regardless of the air," he answered, and then added suddenly, "but it will make a beautiful experiment to drop out a bottle now."

He ceased pumping, and opening a bottle of asparagus tips, he placed them in a bowl, and prepared to drop out the bottle. I took my pencil and wrote this message to go inside,—"Behold, I have decreed a judgment upon the Earth; for it shall rain pickle bottles and biscuit tins for the period of forty days, because of the wickedness of the world, unless she repent!" And I pictured to myself the perplexity of the poor devil who should see this message come straight down from heaven!

In order to make his experiment more successful, the doctor put in half a dozen bullets from one of the rifles, to make the weight more perceptible. Then he put the bottle into the discharging cylinder, and preparing to push it out he stooped over the port-hole. At a signal from him I gave the pump handle several quick, successive motions, and at the same instant he let drop the bottle. At once he cried out,—

"Beautiful! and just as I thought."

"But I didn't see it!" I protested. "What was it?"

"The instant the bottle was released the discharged air was immediately attracted toward it, and gradually surrounded it entirely. It was like a little planet with an atmosphere of its own, as they fell back to the Earth together."

"But I couldn't see it; I had to pump," I complained. "We must do it again."

"We shall soon have our bottled things all emptied out on plates to dry up and spoil," he objected. So I emptied a biscuit tin this time, and delaying for no message, I put it in the discharging cylinder. Then I bent over the port-hole and gave the signal for the pumping. As I thrust out the tin I was astonished to see the lid pop off the first thing. The quick expansion of the air inside it did that. This air, as well as the air from the discharge pipe, seemed to flee from it instead of surrounding it, as the doctor had said. I continued watching so long that he finally said,—

"Hasn't it fallen out of sight yet?"

"No; it is not falling away swiftly as the air does. It is following the projectile! It is not gathering any air about it as you said it would. It does not quite keep up with us; but considering our speed, it is doing remarkably well!"

The doctor was not inclined to believe me until he had looked for himself. He watched and pondered for a minute or two. Then his surprise ceased, and he spoke in that assured way which always irritated me.

"Quite natural, after all," he said. "That biscuit can is made of thin sheet-iron with a surface coating of tin. The iron has become magnetized by induction, and the Earth repels the can just as it repels us. It will follow us to the dead-line, and probably on to Mars, unless the sheet-iron loses its polarization. If we had cast out a thing of solid iron, it would rush ahead of us, instead of falling a little behind, as this does, for it would have no dead weight to carry. But we could not put such a thing out of the rear end, for no force would make it fall that way. If we put it out of the forward port-hole, it would beat us in the race toward Mars."

I remarked to the doctor that the air-pump seemed to be incorrectly built, for its action was strangely difficult in the reverse manner that it should have been. The down strokes went by themselves with a quick snap, but the up strokes were as if against pressure, and the moment the handle was released it flew down again. He had not tested the pump at the surface, as it was of a well-known make, but it certainly seemed to work backwards. Moreover, the more nearly we had a compartment emptied of air, the more difficult the pumping should become, but here again the reverse seemed to be the case, for the longer we worked the easier the up strokes became.

The temperature of the projectile was still fairly comfortable, and the doctor allowed the condensed air to issue very slowly into the partial vacuum in his compartment until it produced a barometric pressure of twenty-seven. Then we pulled back the bulkhead, and when the new atmosphere had mixed with the old in my compartment, a pressure of twenty-eight resulted.

"That is about the way the barometer stands during tempests at sea," remarked the doctor. I could not notice much difference from the air we had previously had. Possibly it was fresher and slightly more exhilarating.

The effort at the pump had made us both hungry again, and I prepared from meat extracts a warm and rather thick gravy to put over the asparagus tips. I attempted to pour it, but it was so light that its sticky consistency prevented it from running. We had a hundred such examples daily of the changes which lack of weight caused in the simplest operations. With sandwiches made of biscuits and condensed meat, we eked out a luncheon. This must have been about noon, for when it was over I remember noticing that we no longer needed the gas in the compartment, for there was a gradually increasing mellow light outside.

"Are we already emerging from the shadow?" I inquired eagerly.

"No, not yet," replied the doctor. "But we are

now entering its illuminated core. I must prepare to photograph the strange appearance of the Sun that we shall see presently."

I hastened to the port-hole, and did not leave until it was all over. What I then saw was one of the most beautiful things of the whole trip. The light outside was not bright, but soft and dreamy, like the first twilight after a rich day of summer. The great corona all around the outer edge of the Earth was the most magnificent appearance I have ever seen. It was not at all dazzling, but had the melting shades, first of a sunrise and then of a gorgeous sunset. We had missed the gradual appearance of the phenomenon, but we had a good view of its highest splendour. The colours were continually but slowly changing, and finally the darker hues gradually suffused and dyed the pinks and crimsons.

The Earth was now about three times the diameter of a rising Full Moon, and the corona was about a quarter her width, and looked as if twenty shell-pink suns were set one against the other and overlapping all about the edge of the dark orb.

"How do you know that is not really the extending edge of the Sun?" I asked the doctor. "Perhaps we are already far enough away to see it all about the Earth like that."

"If that were really the Sun, the light from his extending edge would illuminate the surface of the Earth towards us. The planet's outline would be irregular and partly glowing, but you see it is quite dull and dark, and the outline is most plainly visible."

In rapt attention I watched the delicate shell-pink change to a deeper hue of orange, and then our twilight waned a little and turned a sombre grey. Presently the corona glowed a rich maroon, gradually dying to a luminous purple, which slowly deepened and darkened, and finally melted into the general blackness. And lo! we were in the shadow again, and the dreamily beautiful panorama was over.

"It must have lasted nearly an hour," said the doctor. "I am sorry we did not notice the beginning, but it must have commenced with the same dull shades we saw at the end, and gradually changed to brighter colours. I secured three negatives when the glow was most intense."

"Then we have had a waxing and a waning twilight coming together in the middle of our night. And the corona was like a sunrise, followed immediately by a sunset," I exclaimed.

"And why shouldn't it appear so?" said the matter-of-fact doctor. "Twilight is the commonest phenomenon of refraction with which we are acquainted, and sunrise and sunset are merely a mixture of refraction and reflection. There is nothing new about it."

"Now, Doctor, we must remain friends, but you shall not continually tarnish my poetry with your accursed science! I thank my Creator that He made me ignorant enough to admire the beauties of nature. You are continually peeping behind the scenes, and pointing out the grease paints, the lime-lights and the sham effects. Let me enjoy the beauty of the tableau, no matter how it is produced. I would give all of your pat knowledge for that feeling of profound awe which rises in the untutored breast at beholding the magnificent grandeur of unfamiliar nature."

"When your ecstasy has quite passed, I shall appreciate a little cold mutton and biscuits, and then we must pump out again," he replied.

CHAPTER XI

Telling the Time by Geography

AFTER supper I went up into his compartment, and having arranged the bulkhead, began the tedious operation at the pump handle. It was a matter of pure muscular strength, as the effort had to be made to lift the handle, which snapped back sharply when released. I was working vigorously when I was suddenly struck dumb at seeing the handle break off just at the point of leverage, so that it was quite impossible to operate it. The doctor heard the handle fall, and looked around in great vexation.

"That means asphyxiation within twenty-four hours!" he exclaimed.

"Which is plenty of time to think it over," I answered.

After all, why was this pumping necessary? If a way could be devised to open a valve, all the air would rush out of my compartment as easily as beer runs out of a bung-hole. In fact, it did rush out a little at a time, which is what made the handle go down of itself. But any such new valve would have to be automatically closed, as it would be manifestly impossible to enter and shut it. I kept on thinking, and finally began examining the partition between the compartments. There seemed to be several long screws that went quite through it.

"Doctor, did you ever hear of those wise people who, after every freshet, shipped the surplus water down the river in boats? Well, it strikes me this air-pumping is just about as useless labour. Help me pull in the bulkhead and I will show you something."

I went at once to the cylinder we used for discharging things from the projectile. With a pair of pliers I chipped off a small piece of the edge of the closing lid in two places, one near each end. This made two little irregular holes into the cylinder about eight inches apart. Then I pushed it half way out, so that one hole was outside and the other inside. Of course the air rushed through the inner hole into the cylinder, and thence through the outer hole to the exterior.

"Shut that thing!" cried the doctor, when he saw what I had done. "Do you wish to suffocate us? That will let the air out perfectly, but how are you going to close it to admit the condensed air?"

"People unskilled in these matters are so hasty!" I said rather sarcastically. "Wait until I have finished and you will see."

I found he had a screw-driver, and I loosened one

of the long screws and enlarged the half of its hole toward my compartment. Then I whittled a block of soft wood, so that it would slide smoothly into this half of the hole. Driving the screw home again, I just allowed its tip to enter the end of the block. Then I fastened a piece of stout twine to the cylinder and the other end to the block of wood, which was almost opposite it. Pushing the cylinder half way out, I made the twine taut, and hastening into the doctor's compartment, I thrust in the bulkhead. The air was rapidly escaping. Waiting long enough for all of it to have leaked out, I then unscrewed the long screw, which gradually drew in the block of wood and the twine, and thus pulled the cylinder into the projectile so that there was no connection with the exterior. Then the doctor let in the condensed air to a barometric pressure of twenty-six, and the whole operation was over in a few minutes. My compartment must have been almost a complete vacuum. When it was over, I cried rather triumphantly to the doctor,—

"There, you see, one doesn't need a steam pump to make the water run over Niagara! At this distance from the surface, nature abhors a gas and prefers a vacuum!" He was inclined to be rather sulky at first, but he really did not like pumping any better than I did.

I should say it was about five hours later that we noticed it was growing gradually lighter outside.

Mars lost his ruddiness and grew pale in a grey field. Our view of the Earth was also becoming more and more misty.

"We are emerging from the black core of the shadow into the semi-illuminated penumbra," said the doctor. Then he altered his course experimentally, and found a slightly darker path, but it soon began changing again to grey.

"There is no use trying to keep in the umbra any longer. It is growing too narrow. The penumbra will last quite a long time yet, but it will gradually get fainter and fainter. We shall not plunge at once into the dreadful light you fear so much. Keep your eyes glued to the Earth. I can scarcely see Mars any longer. The whole field is getting blank and white."

The rear vista was also growing a pale white, and I could distinguish the form of the Earth as a darker object slightly larger than a full moon when risen. But it was all growing dimmer and dimmer as the penumbra faded toward the perfect light.

"Mars is completely gone now," said the doctor. "The field of the telescope is one pale curtain of light. I have steered to the left to go ahead of him now, as there is no longer any reason for going behind him."

I heard him working at the telescope as if loosening it from its fastenings, but I dared not take my eyes from the Earth to see what he was doing. Presently he called out to me,—

"Make room down there. I must bring the instrument down and observe the Earth now. Be careful you don't lose sight of her." But the instant he removed the telescope from its bearings and uncovered his forward window, I lost all view of the Earth. The new light now entering by his window, from behind me, made it impossible to see so far.

"Too late!" I cried; "I have lost her! We are alone in limitless space, without even the company of the planets!"

But while the doctor was carefully lowering the telescope, my eyes were still searching, and presently I perceived a thin crescent of faintly brighter light, growing gradually wider. It was like a new moon dimly seen in a clear part of the sky when the afternoon sun is cloud-hidden. The doctor stopped to look where I pointed it out to him, and then changed the wheel a little.

"That is a thin slice of the illuminated part of the Earth," he said. "We can no longer see the dark side which has been visible to us while in the shadow. Fortunately our new course a little ahead of Mars will give us a constant view of this thin crescent."

We now stood the instrument on end over the port-hole window, which brought the small end near the aperture between the compartments. When the doctor had secured a focus, he called me to look. The crescent was greatly magnified, but the outline of the sphere on the other side could not be seen, nor could anything be distinguished in the centre. Both the outer and inner edges of the crescent were ragged and irregular in places, and there were faint darker spots on its surface. I called the doctor's attention to the fact that the ragged appearance was always in the form of extending teeth on the outer side of the crescent, and in the form of notches eaten into its inner edge. He studied all these appearances carefully and finally said,—

"This crescent is that part of Earth which is just coming into morning. It is gradually shifting from east to west with the Earth's rotation of course. What we see now, however, is *land* almost from pole to pole. There is a small sea just above the middle, which might be the Mediterranean. Moreover, it must be mountainous land to cause the ragged edges and the shadows inside."

Then he turned away to get his globe, and I took the place at the instrument. He was slowly turning the globe and examining it thoughtfully as he said to himself,—

"The only continuous land from pole to pole with one interrupting sea must be over the two Americas or over Europe and Africa. The American mountain ranges run from north to south, while through Europe and Africa they are scarce, and almost uniformly run from east to west. Besides, the sand of Sahara would be sure to show as a large, bright, regular spot. A section from longitude 70 to 80 west would include the Green Mountains and the Alleghanies of North America and the Andes of South America, and in that case the darker spot in the centre would be the Caribbean Sea."

"Look here!" I cried. "Toward the lower end the inner outline is growing darker but more regular, and faint streaks or shadows reach through the brighter light toward the dark greenish regular surface which looks like water."

He observed closely and said,—

"Those shadows must be cast to westward by the enormous peaks of the Andes, and the dark greenish surface they reach toward must be the Pacific Ocean."

Then he consulted his globe while I looked. "The first two to come into view," he said, "would be the two great peaks in Bolivia, over twenty-one thousand feet high."

"There *are* two of them together," I said, "and now others are rapidly coming into view. There are five more scattered unequally, and then, lower down, three near together."

"Then there is not the slightest doubt that we see the Lower Andes," he said. "These last you mention are scattered just as you say along the border between Chili and Argentina, and the group of three are near Valparaiso, the peak of Aconcagua being the tallest. But watch now for the group in Ecuador, about midway between the top and bottom of the crescent. There are four very large peaks and numerous smaller ones."

"The middle all looks bright yet, like land, with no shadows or greenish spots. But a queer thing is happening lower down, where the shadows have ceased lengthening and are now fading. There are several fine points of light just beyond the outer edge of the crescent. They are mere bright specks, but gradually they join with the surface, making a rough toothed edge."

"Ah, that phenomenon has been observed upon the Moon," said he. "That is the sun shining on the snow-capped peaks first, and then, when the diminutive outline of the mountain comes into view, it looks like a tooth."

"The same is happening all down the coast," I reported. "Now I see it on the lower group of three."

"Give me the instrument," demanded the doctor. "That can be nothing but the west coast of South America, and if that be the case, the whole thing will be repeated for the tall group in Ecuador, dominated by Chimborazo."

As I surrendered the telescope to him, the whole lower part of the crescent was dark, but with regular edges. Only in the middle, which should have been about the Equator, and in the upper part, was there the bright lustre of land reflection. He watched for fully half an hour before observing anything remarkable. At last he exclaimed,—

"Now they are beginning! Five streaks near together and just at the Equator. They are almost equidistant from each other, and the next to the lowest one is the longest. Now the top one begins to fade! Yes, and a point of light has appeared detached from the outer edge, and now another and another! They are growing inward toward the surface. Now they are all connected like five saw teeth; the bottom one is the shortest, and that next very high one is old Chimborazo."

"Then it is morning at Quito and also at Pittsburg!" I said, tracing up the 80th meridian.

"Yes, and we have been one complete day and about five hours more travelling the nine hundred thousand miles that lie between this and Earth," replied he.

"That makes us one full meal behind time," I said; "but we have discovered a way to make the Andes call us for breakfast. When the Pacific Ocean has passed from view, Japan and Australia shall strike noon for us, and we will have supper and call it night when the Indian Ocean is gone and darkest Africa has come into view!"

CHAPTER XII

Space Fever

WE counted seven successive returns of the peaks of the Andes, and being by that time certainly six million miles from the Earth, we could distinguish them no longer. Then followed what I remember as a very long and unspeakably monotonous period, without any adequate method of marking the time. Our days became a full week long, for the only way we could guess at the time was by the quarterings of the Moon. We could still see her about the size of a marble in the telescope, and as her crescent began to wane, and finally her light entirely disappeared, we knew she was then just between us and the Earth, and shining upon that planet as a Full Moon. This was due to occur fifteen days after our departure. Then we watched her grow from a thin crescent to a bright quarter, and we knew another week had elapsed.

"We shall soon be able to determine one date with absolute certainty," I said to the doctor, when we must have been some twenty days out. "I have been reading up your almanack, and I find there is a total eclipse of the Sun by the Moon on June 29th."

"You might as well try to eclipse him with a straw-hat, as far as we are concerned," he replied. "The Moon will necessarily be on the further side of the Earth when that occurs, and the eclipse will barely reach the Earth. It will fall short of us by a matter of some thirty million miles!"

It was soon after this that we gave up observing the Earth as a planet, put on our darkened lens, and proceeded to hold her as a spot in the Sun a little to the left of his centre. The Moon remained a tiny spot of light outside for a few days; but finally she entered the Sun also, and was seen as a faint spot travelling toward the Earth-spot.

Although the dazzling quality of the light, into which we had emerged after the second day, was finally beginning to wane and pale a little, Mars was still invisible. In fact, no stars or planets were visible; only the gleaming Sun with the Earth-spot upon it. Our thermometer was poorly placed in the glare of the Sun at the rear; but it showed the heat was decreasing, and from a temperature of thirty-five degrees, observed at the end of the second day, it had now fallen to twelve, and was diminishing regularly about two degrees daily as nearly as we could reckon.

Our appetites were steadily failing, and for two very good reasons: the unsuitable foods and the impossibility of getting any exercise. There was no such thing as getting any healthy actions of the body. Nothing had any weight, and such a thing as physical labour was impossible on the face of it. I attempted to go through regular courses of gymnastics at frequent intervals; but as my body and its members weighed nothing, my muscles found nothing whatever to expend their force upon. I thought myself worse than Prometheus bound upon his rock, for he could at least struggle with the birds of prey and pull upon his chains! I might as well have been utterly paralyzed, and I actually began to fear that I should lose all my strength, and that my muscles would forget their cunning.

And our foods could not have been more unsuitable. The light vegetable diet which this lack of exercise called for was impossible. We had never had any fresh vegetables or fruits, and our tinned and canned supplies of these had been rapidly exhausted. We had plenty of solid, meaty foods and beef essences; but our systems did not require these, and at last absolutely refused to have them. I lived for days at a time upon beer and biscuits, and looked longingly at my cigars. I believed I could have existed comfortably and luxuriously upon smoke alone. My dreams were filled with visions of ripe, luscious fruits and fresh, crisp vegetables. When I awoke, I loathed the only foods we had.

I believe I should finally have given up eating,

had I not hit upon a method of exercise at last. It was a sort of rowing or pulling machine, which I rigged up by running a bar through one end of the doctor's spring scales, and fastening the other end to the foot of my bed. I pulled vigorously against this spring for hours at a time, and was delighted to find that my strength had not left me, and that I could easily lift as much as these scales had been made to weigh. I remember the returned appetite with which I enjoyed potted meat and a tinned pudding, after the first hour of as vigorous exercise as our rarefied air would permit.

The Moon-spot had disappeared and gone to her eclipse behind the Earth, when an incident occurred to vary the monotony of our existence a little, and to suggest to me a diversion that had been hitherto forbidden. Our supply of water in the outer tank had long ago boiled away, and I had lighted the gas to heat water for the doctor's coffee. I had taken the cup up to him and remained chatting with him, when presently I smelled something burning from the compartment below. I descended quickly, and saw that my light bedclothes, which now weighed less than a feather, and often floated from their place, had been drawn into the flame by the draft of the burning gas. They were floating about the compartment now, all aflame and threatening to set fire to everything. We had not a drop of water to spare; but for once I thought of the right thing to do without hesitation. I pushed out the

I

ventilating cylinder, hurried back to the doctor's compartment and thrust in the bulkhead. Within two minutes all the air had escaped from my room, and the fire had died for lack of oxygen. I waited a few minutes longer for the smoke to escape, and then we admitted condensed air, but only to the remarkably low pressure of eighteen. Within five minutes the compartment was ready again, and there was not a trace of smoke or smell of fire to be perceived.

"I congratulate you on your quick perception and prompt action," said the doctor when it was over.

"Quick rubbish!" I exclaimed. "I have been a dundering fool for four weeks by the Moon! I might just as well have been smoking ever since I contrived this self-ventilating arrangement. The compartment becomes a perfectly clean vacuum at each operation, yet I had to wait for this bed clothing to catch fire before I could think of so simple a thing!"

It was at the meal time just preceding the next changing of air that I opened the last tin of canned peas, as a sort of treat for the doctor to offset my expected revel in fragrant tobacco. I prepared half the quantity for him, but left my portion in the tin until I should be hungrier. With the prospects of a good smoke before me, I had no appetite for food. I put in the bulkhead to prevent the smoke from entering his compartment and lighted my Havana.

Then I took Two-spot on my lap and stretched myself for a reverie. On Earth, smoking time had been my period for reflection. And far back on that distant planet, what were they doing now? In that one busy corner that had known me, they had probably wondered at my disappearance for a day or two; but after the month that had passed I was certainly forgotten. There were few back there whom I cared for, and not many had much reason to remember me. My interests, my desires, my hopes were all ahead of me on a new planet. And what was waiting for me on Mars? Discovery, riches perhaps, and a measure of fame when I returned. Then I thought of the numberless problems that the next few weeks must solve for us. Would there be intelligent inhabitants on Mars? Would they be in the forms of men or beasts? Would they be civilized or savage? Would they speak a language, and how could we learn to communicate with them? Would they have foods suitable to us; indeed, would the very air they breathed be fit to sustain our lives? Should we find them peaceable, or, if warlike, should we be able to cope with them?

These thoughts were interrupted by the doctor, who called feebly to me to come up. "Don't eat any of the peas," he said weakly. "There was a queer taste about them, and they have made me deathly sick."

He was very wretched, and grew rapidly worse.

I immediately saw that it was a severe case of poisoning, and I did everything I could to relieve him, but he groaned in agony for several hours. Finally he fell asleep, but his rest was disturbed by fits of delirium, in which he raved wildly in German mixed with English. As he slept I had time to think the matter over carefully. After all, it was a thing which required only simple remedies, and I had administered them. It was only a question of a little nursing and a careful diet, and he would be well again.

But his fever increased and his delirium became more frequent, and I began to appreciate that the derangement incident to the poisoning had prepared the way for a more serious illness. During his ravings I caught a glimpse of the struggling and ambitious side of his nature, which he always so carefully repressed.

Once I heard him mumble this to himself in German: "Kepler perceived a little, he saw dimly; Newton comprehended the easy half; but Anderwelt, Anderwelt of Heidelberg, grasped the hidden meaning!"

In spite of all my attentions (I did not then understand the nature of Space Fever, of course), he was growing steadily worse, and I was becoming desperate. I could not afford to have him ill long. The currents would probably continue to work fairly well until it became necessary to reverse them, and that time was not far off. Unless they were reversed exactly at the right moment, we might fall into the neutral spot and be held there for ever. Even if I managed to stop the negative current, and succeeded in falling towards Mars, I could not regulate the positive current so as to temper our fall and make a safe landing. It was equally dangerous to remain fixed in space, or to fall headlong upon a planet and be smashed, or be buried miles deep if the projectile did not collapse.

I had no way of telling how much time passed, but it seemed to me a very long period, and he grew steadily worse as we approached the neutral point. I tried to rouse him from his delirium. I addressed him jocularly, then commandingly, then beseechingly. And he answered me always with reflections from that other side of his nature which one rarely saw when he was well.

"Hast thou seen red ants crawling upon a cherry? Such are the mere circumnavigators of a globe! What! Hath not the world forgotten a Columbus? How long, then, will it remember —— Hast thou no cooler water? This is tepid and bitter!"

Ever since the last quarter of the Moon, which must have been ten days ago, there had not been the slightest perceptible evidence of movement. The standards by which we judge motion on the Earth had failed ever since we left the atmosphere. There was no rushing or whizzing; we passed nothing; all the ordinary evidences of speed were absent.

When you lie in the state-room of a smoothly moving steamer, no forward motion is perceptible. If you see another ship pass near by, you get a sudden surprising idea of the speed. If you watch the receding water, you appear to be going forward slowly; and if you watch the spray at the bow or the wake astern, you appreciate the movement more fully. But if the waves or the tide happen to be running with the ship, she has apparently almost stopped, when really her speed has been somewhat accelerated. If you watch the distant stars, you can scarcely perceive any motion at all; and if the clouds should be moving in the same direction as the ship, her motion appears reversed.

We had none of these things by which to judge, and we appeared to be hanging perfectly still in space, though the doctor had assured me we were travelling at least five hundred miles a minute. This was rational, as it agreed with the diminishing size of the Earth; but it required an effort of faith on my part to believe that we had been moving at all.

But suppose we should gradually lose our speed and stop in a neutral point, how should I know it? The Earth now was, and had been for ten days, a mere spot on the Sun. While Mars had been visible, he had never increased in size in the telescope, and he was now invisible. The only way I could tell would be to wait until after many days had elapsed, and if Mars did not finally come into view, I should know something was wrong. But it would be too late then; there would be no winds or tides, no weight or buoyancy, nothing to move us out of that dreadful calm where even gravity does not exist. That must be avoided at every cost! But might we not be very near it now? Weight had been practically nothing for a month, within an hour it might be positively nothing, and——

The doctor's mutterings interrupted these thoughts. "The power with which to travel was so simple and so vast! It all lay hidden in that elementary law of magnetism, like poles repel and unlike poles attract. But the road to travel and the problems by the way, those were the hard things!"

He was putting them all in the past tense, as if he had already solved them! But what was that law of magnetism he mentioned? Perhaps he would reveal his secrets to me in his ravings! I must mark every word he said; for it was clear I must solve the problem, he would not be well in time. I must brush the cobwebs from my meagre science and struggle with his invention.

"Unlike poles attract," he had said. Then Earth and matter must normally have unlike poles, and to make Earth repel matter it would only be necessary to change the polarization of the matter. Yes, he had told me it was all accomplished by polarizing the steel and iron of the projectile!

When they were made the same pole as the Earth, then she repelled them. But if the whole thing were so simple, why had it never been discovered before? Ah, that is the strong shield behind which incredulity always takes refuge!

I ventured near the gravity apparatus and examined it carefully. There was a small thing which looked like the switchboard of a telegraph office. The perforations in it were all in a row, and the ten holes were now filled with little brass pegs, which were suspended from above on small spiral springs. These were evidently the points of communication of the negative current to the framework of the projectile. It certainly would do no harm to pull out one of these pegs, as that would only slightly diminish the current. At least I would risk it. My fingers had scarcely closed upon the brass, when I was given such a violent shock as to be thrown powerfully across the compartment; and had my body weighed anything, my bones would certainly have been broken by the concussion. My arm and shoulder did not recover from the stinging and deadening sensation for some time. I noted the little peg I had pulled out hanging by its spiral spring just above the hole it had filled. It would be worth my life to remove the other nine in the same way.

Besides, how would I know when the time came to remove them? My eyes fell upon the two large leaden balls suspended from short copper chains.

I had seen these before, but now I thought I understood them. They would swing whichever way gravity attracted. They hung down toward my compartment now, and if we ever passed the dead line, they would hang forward toward Mars. But in the neutral point what would they do? When the gravity of planets neutralized each other, the steel of the projectile would repel these balls towards its centre, which would tend to put them both in the same spot and thus bring them together. Moreover, they would slightly attract each other. Yes, it was quite certain that these had been devised as a Gravity Indicator, and they would tell me when we were approaching a dead line, when we were in it, and when it was safely passed. But all that would do me but little good unless I could manage the currents.

I sat thinking this over a long time, when it suddenly occurred to me that the doctor would recognise, even in his delirium, the importance of action when these two balls came together. As soon as they had approached each other, I must lift him up and show them to him. The brain that had made them would know their meaning, and know how to act even in illness! Perhaps I was like a drowning man clutching at a straw; but from the moment I thought of this I believed firmly that the solution of the whole problem would come in this manner. My hopes were ready to hang on the slightest peg. It consoled me to remember some instances where men temporarily insane had been brought to consciousness by impending danger, or by the sight of what last weighed upon their mind.

When I glanced at the balls next, I saw that their chains lacked an inch of being parallel. They were already moving slowly inward toward each other. I noted that the chains, which ran through the balls and were connected with a small copper plate on the bottom of each, were just long enough to allow the bottom edges to touch, if they were drawn as far toward each other as possible.

The doctor's fever was at its very worst, but that did not dampen my hopes. The balls were gradually drawing nearer together. I wished them to be quite close before I made the supreme trial which was to liberate us or leave us prisoners in space for ever! Presently I loosened the knotted sheets which held him to his bed, and lifted the feverish man, as I might have carried a doll, and brought him in full view of the approaching balls.

"Doctor, listen now and look," I said firmly and commandingly.

"Always stubborn and unbelieving!" he raved. "I must take it to a new country, to America, where they invent things themselves, and are willing to listen, and anxious to try!"

"Doctor, don't you know me? It is I, Werner, who helped you. This is a crisis for us! Do you see those approaching balls? You know what they mean! You must save us."

"Thou'rt too busy, like all the rest! Why, then, remember that to-morrow will despise those who are so busy with to-day! Opportunity has knocked and listened for thee and thou hast bade her begone!"

"Listen, Doctor. I am he who heard you and gave you the pink cheque. I am he who refused three times to go with you and then came at last. I am he who was afraid of the light, who dodged the Moon, and chaffed you about the pump. Do you not remember it all? Come, you are no longer ill. There is work to do. Have you forgotten the leaden balls? See! they are touching each other now, and we are in the dead-line, the neutral spot, the one danger of the trip which you acknowledged."

But it was useless. He remembered nothing, his eyes were dim and vacant, and the great brain that had planned all this was overthrown by fever. The experiment had failed and we were lost!

I tied him gently back on his bed and turned in desperation to the apparatus, deciding to risk my life to pull out those nine pegs with my hands, one after another.

My God! they were already out! Every one of them was hanging by its spiral spring, just above the hole it had filled. The switch-board had opened a little and released them. It was all automatic! The contact of the copper surface of the balls had completed a short circuit which cut the negative current. He had thought of it all, even to this emergency, and the machine could take care of itself!

And in the wave of thankfulness and rejoicing which swept over me, I sank on my knees and kissed the forehead of the feverish old man again and again!

CHAPTER XIII

The Mystery of a Minus Weight

IT was the doctor himself who gave the name Space Fever (now so generally adopted) to the peculiar malady from which he suffered in that long period when weight was very slight or nothing at all. A little reflection on the physiological bearings of the conditions we were passing through, will serve to explain the illness.

For the period of a month, owing to the impossibility of effort, there was scarcely any wasting of our bodily tissues, and very little need for oxydization of the blood. The limbs, which the heart really works hardest to serve, did scarcely any labour and needed very little blood. But the heart had its stubborn habits the same as the other muscles. It is a high-pressure engine, and there is no way of slowing it down materially. It kept up its vigorous pumping and driving just as if the great muscles of the limbs had wasted and needed building up, and just as if it had the task of forcing the blood through those parts of the body usually compressed by its weight or strained by the effort of carrying it. The result was much the same as if your heart now should suddenly begin to beat much too fast, the blood was heated into a state of fever, which naturally increased as we lost weight, culminated at the dead-line and began decreasing as soon as we commenced having a weight toward Mars. It was only my fortunate invention of a method of exercise, and my religious adherence to it, which saved me from a similar attack.

But many things happened before the doctor recovered consciousness. The Moon had reappeared on the other side of the Earth-spot, the light about us had grown less dazzling than sunlight on Earth, and the temperature had fallen to four degrees. It was perhaps two days after passing the dead-line that, as I was gazing carefully out of the forward window, I saw far to the right of us a large circular patch of faintly redder light in the general curtain of white. Its size quite startled me, for it was rather larger than a full moon, and I had expected Mars to re-appear as a very bright star before we could distinguish any disc with the naked eye. This misapprehension probably arose from the fact that I had thought the dead-line about half way between the two planets, which upon reflection I saw to be impossible, as it must be much nearer the smaller planet.

The outline of the planet was not clearly visible yet, but I could not have missed seeing that red

THE MYSTERY OF A MINUS WEIGHT 143

glow long before, had it been more directly in front of us. Evidently we were steering much ahead of the planet, which indicated that we were arriving before opposition. I immediately changed our course so as to go more nearly toward it, but yet to keep a little ahead. Then I hastily brought the telescope back to the forward compartment, which was now the bottom of the projectile. The lenses easily pierced the curtain of light that seemed to be hung in front of the new planet, and I could distinguish the outline of the greatly magnified orb very clearly.

Judging from appearances, it could not be farther from us than twice the distance of the Moon from the Earth. I resorted to the scales at once, and found that weight was beginning slowly to return, for I weighed a little less than an ounce. From a rule the doctor had explained to me, I calculated that this indicated a distance from the planet of about four hundred thousand miles, if it really was Mars. But I had some doubts about its really being that planet; for a clear white, irregular-shaped spot upon it, which I had noticed as soon as the telescope was focussed, did not appear to move at all, as it should have done had it been upon a rotating planet. Upon closer observation, I detected a dull, greenish spot, just coming upon the lower edge. But when I looked again a bright white and perfectly circular spot had appeared in the same place and covered it up. But this new white spot travelled much more rapidly, and soon uncovered the greenish spot, which seemed to move in the same path, but much more slowly. This was something I could not understand. The white circle was too bright and regular to be a cloud, yet if they were both on the surface how could one travel faster over the same path?

Very soon the white circle passed entirely across the greater orb, and then I was surprised to see it detach itself from the planet and remain for a few moments as a separate small orb in the sky! Could this be another freak of refraction? But before I could determine, the little orb disappeared behind the greater disc and was gone. The greenish spot, which I judged to be truly on the surface and caused by an ocean or great sea, was about three times as long in crossing the disc. I next turned my attention to the immovable and irregular white spot, and discovered that its edges seemed to be revolving slowly around its centre. Then it occurred to me that this spot must be located at one of the poles and be caused by polar ice and snows. The doctor had expected such on Mars, and I no longer doubted that this was our objective planet.

It was like a great holiday for me when the doctor regained consciousness. Almost as soon as his fever abated he was well enough to perform his customary duties. His illness had not made him appreciably weak, because as yet scarcely any effort was required to move about. He was quite as anxious to hear all my experiences as I was eager to relate them. I gave him a full account of my struggle passing the dead-line, of my discovery of Mars, and the various spots I had noted.

"From the time it took the greenish spot to cross, I should judge a Martian day to be about fifty hours long," I said.

"Then you *must* have been very lonely," he replied. "For a Martian day is just forty-one minutes longer than an Earthly day, unless a great number of our scientists have continually made the same mistake in observing him."

"When we arrive, we shall be able to determine the point exactly if our watches commence running again," I answered. "But I think I know one reason why I have misjudged the time. Ever since you have been ill I have slept very little. I have hardly felt the need of rest since I lost my weight. I have been growing more and more wakeful, and I rarely sleep more than an hour at a time. That seems quite sufficient to refresh me."

"As we regain our weight we shall feel the need of sleep again," he said. "But on Mars we may need but one-third as much as we had on Earth, unless we exert ourselves proportionately more."

Then I told him about the circular spot which had seemed to slip off the upper edge of Mars, and asked his explanation of it.

"That must have been Phobos, one of the moons of Mars," he said.

K

"One of his moons!" I exclaimed; "I didn't know he had *any*."

"You are an American, and say that!" he answered in surprise. "It is one of the astronomical glories of your people that they discovered the two moons of Mars, during the favourable opposition of 1877."

"This is the first case I remember where we have left it to a foreigner to tell us how great we have been!" I laughed.

"These two moons of Mars also furnish a most interesting example of how fiction may forestall and pre-figure actual scientific discovery. Dr. Swift made Gulliver, in his wonderful travels, discover two moons of Mars, revolving at a speed which he must have thought ridiculously fast. Many years afterward the American telescopes really found two moons, but actually revolving more rapidly than Dr. Swift had dared to boast! If your white circle was really Phobos, you have seen the freak among satellites. She is the smallest, swiftest moon ever discovered, and travels so much more swiftly than the revolution of her primary that she appears to go opposite to everything else in the Martian sky, rising where the Sun sets and crossing the heavens from west to east!"

"What I saw did travel in the same direction as the rotation of the planet, and much more rapidly," I exclaimed.

"Then it was Phobos without a doubt, and she is

THE MYSTERY OF A MINUS WEIGHT 147

due to appear again in the west in three hours and fifty minutes after she sets in the east. We must watch closely, for I wish to land upon her and make a flying trip all around Mars with her. Do you realize what a glorious view we shall have of the great planet, sailing around him on this satellite in a period of a little over seven and a half hours, and at a distance of only about four thousand miles? There will be no night, for if one side of the little moon is heavier than the other, the heavier side will always be turned toward Mars. Therefore, when the Sun does not shine on Phobos, Mars will do so, and keep her continually illuminated, except for the brief period of the regular eclipse during each revolution. And one-fourth of the entire heavens, as seen from Phobos, will be filled with the glowing orb of Mars! The great planet will exhibit to us at a near range all the configurations of his surface, his oceans and his clouds. We will survey and photograph him to our hearts' content."

The doctor was justly enthusiastic on this subject, and I felt that such a landing would, in some measure, compensate for my disappointment in not being able to visit the Moon.

As I watched carefully, the satellite finally came into view, but very much more distant from Mars than before. Also, it moved very slowly now, and seemed to grow larger as it approached the disc. I pointed it out to the doctor, and remarked that it was acting quite differently. Just as it entered upon the orb of Mars, another moon, somewhat smaller, mounted hurriedly from the under side of the planet and began hastily ploughing her way over the ruddy disc.

"That last one is the one I saw before, that is my Phobos!" I cried excitedly.

"Then the other slow one is Deimos, the outer moon. She appears the larger to us now, because her greater distance from Mars makes her nearer to us, but she appears to the Martians as the smaller. We must observe closely, and we may discover some new and lesser satellites which Earthly telescopes have never found."

"Time enough for that when we land on Mars," I answered. "If we get in past these two without being hit, I shall be satisfied. You dare not venture in front of that Phobos, and I don't see how you can ever overtake her if you approach from behind."

"That reminds me to slacken speed, for we must be getting very near," he said. "Please weigh yourself every few minutes and note your increasing weight. You should weigh seventy-two pounds on Mars, and eight pounds at the distance of Phobos."

He immediately reversed currents, and when I reported that I weighed almost a pound, it frightened him, and he turned in the full power of the negative currents to overcome our momentum. And it

THE MYSTERY OF A MINUS WEIGHT 149

proved that the repelling power of Mars at the distance of 15,000 miles, which this indicated, was not at all strong against the great velocity we had been daily acquiring. I hung upon the scales every few minutes, and reported a steadily increasing weight up to three pounds.

"That shows a distance of eight thousand miles," he figured. "Almost exactly in the orbit of Deimos, but she has safely passed, and will not return for thirty hours. We must turn the rudder hard over to the right, and sail around the planet in a circle until Phobos overtakes us; then, if we approach her travelling in the same direction at almost the same rate of speed, her gravitational attraction will pick us up and draw us safely ashore."

Mars was already an enormous orb ahead of us, and many of his features, such as oceans, ice-caps, and continents, could easily be distinguished; but we paid little attention to them, being occupied with making a safe landing on Phobos, and expecting to make a systematic study of him from there.

"We must not attempt a landing on the outer side of the satellite," the doctor reflected, "for we should have no way of getting around to the inner side to make our observations. We must go within her orbit, and then as she comes past allow her attraction to draw us gently toward her."

We had quickly overtaken and passed Deimos, far within her orbit. I was keeping a close watch for Phobos out of the rear window as we circled about Mars at a distance which we calculated, from my weight on the scales, must be within the path of the satellite. We were circling in the same direction that the great planet was rotating, and yet we passed by things on his surface, which proved that we were travelling faster than his rotation. The doctor noticed, with his telescope, a brilliant snow-capped peak of a great mountain towering up from a small island. The contrast of the snow peak, with the darkish green waters all around it, was the most pronounced thing visible on the great planet, and he decided this must be the white spot detached from the polar ice which our astronomers have frequently observed at about twenty-five degrees south latitude, and to which they have given the name Hall's Island.

"I am afraid we have not appreciated the speed at which we have been travelling," remarked the doctor. "Phobos is very slow in overtaking us;" and he was just beginning to slacken speed still more, when he suddenly cried out,—

"Here she is ahead of us now! We have overtaken her, instead of waiting for her to catch us!"

And, true enough, we were gradually approaching a small brownish mass, feebly illuminated on its outer half by the sun, and more faintly still on its inner half by reflected light from Mars.

And how shall I describe that queer little toy-world which we were gradually overtaking?

THE MYSTERY OF A MINUS WEIGHT 151

Imagine, if you can, a little island, less than a third the size of the Isle of Wight, tossed a few thousand miles into space, and circling there rapidly to avoid falling back upon the greater sphere. Imagine that flying island devoid of soil, of trees or vegetation, of water or air, of everything but barren, uncrumbled, homogeneous rock, and you have some idea of the unadorned desolation of Phobos, into which we were slowly sailing, or falling. There was not even the slightest trace of sand or scraps of rock, such as time must have abraded from even the hardest surfaces, but the reason for this soon became apparent.

The doctor feared steering directly against her as we approached, lest we should land with a crash. We had already reached her and were travelling along her inner side. Although we were very near her, she seemed to have very little attraction for us. Then he turned very much closer, but as soon as the influence of the rudder was released, we seemed to leave her instead of falling upon her as we expected. We were still travelling faster than she was, and had we steered directly against her, we would have crashed and bumped against her protuberances. Still there seemed to be no other way to make a landing. In order to estimate the amount of such a shock, the doctor calculated, from the best information he had of her size and a guess at her density, that she would attract the projectile and its entire load with a force of only two pounds.

That was not enough to cause any very great shock, and he decided to take chances at once, before we had entirely passed her. He turned the rudder hard over toward the satellite, and we came against her with scarcely any crash, but with a bumping and grating that continued until the rudder was eased back. Then, to our great surprise, we did not remain on the surface, but rose from it and sailed inward towards Mars.

"Something wrong here!" exclaimed the doctor. "She has no attraction for us."

"Well, how do you explain this?" I asked. "You say the whole projectile weighs only two pounds toward Phobos, when, just a short time ago, I weighed nearly eight pounds myself on the scales."

"True enough!" he cried; "the gravity of Mars must be dominant." He began figuring rapidly, and then exclaimed: "We weigh one hundred and thirty pounds toward Mars, and only two pounds toward the satellite. Small wonder that we could not make a landing, with Mars pulling us away sixty-five times harder than Phobos attracted us! But this is very strange! I remember no mention of this in any of the astronomical writings, and it is as easily calculable on Earth as it is here. Moreover, this must cause everything that is loose upon Phobos to fall upon Mars. The great planet is tugging at everything the satellite has with a force sixty-five times stronger than her own!"

"Now, I am afraid those figures won't do, Doctor," I put in. "For, if what you say is true, what prevents the whole satellite from tumbling into Mars at once?"

"She would do so were it not for centrifugal force. The speed with which she whirls around the planet must just balance the force with which he attracts her, and thus she is kept in her orbit. But stones and loose things on this side of her centre are attracted more strongly by Mars than they are repelled by the whirling, so they must all have fallen to the planet. That is why the surface was perfectly barren. If Phobos always keeps the same side turned toward Mars, there may be rocks and soil on the outer side, and we could land there with a positive current; but we could not see the great planet, as I had hoped."

"I have had quite enough of this moon-chasing," I said; "let us be off for the large game at once!" and the doctor agreeing, we turned directly toward Mars.

BOOK II
Other World Life

CHAPTER I

Why Mars gives a Red Light

OUR telescope was now pointed exactly at Mars, and we were observing every feature as we approached him. Compared with the illuminated crescent of the Earth, which we had studied when we were observing the Andes, our present view was infinitely vaster and more comprehensive. We were approaching the illuminated side of a planet, whereas we had then been rapidly receding from the dark side of one partly lighted at its edge. In our new vista there were remarkably few clouds. There were a few pale mists here and there over the seas, but no such heavy, black masses as had frequently obscured the Earth.

On Mars there were fewer large bodies of water, and a very much greater proportion of land. In fact, about the Equator, whither we were steering, there seemed to be a broad, uninterrupted zone of land, with occasional bays or inlets cutting into it, but never crossing it. An open sea of considerable proportions surrounded the great ice-cap at each pole, and it was apparently thus possible to travel entirely around the globe, either by sea or by land, as one might choose.

"Behold again the infinite wisdom of the Creator!" cried the doctor. "Although Mars is a much smaller planet than our own, it is fitted for almost as large a population. The land is nearly all grouped about the Equator, where it is warm enough to live comfortably. On the contrary, on Earth there is no important civilization under the Equator, and most of the land is favourably located in the north temperate zone. On Earth the intervention of great oceans between the continents kept the population restricted to Asia and Egypt for centuries, and to the Old World for a still longer time. But here, this band of continuous land has made it easy and natural to explore the whole globe, and its inhabitants have had ample time and opportunity to distribute themselves."

But by far the most wonderful thing that we had been observing for a long time, and which became more remarkable as we approached, was that the entire planet, seas and continents alike, gave off a reddish light. This tinge of red had been visible ever since we had left the Earth. Much further back we had observed that it seemed to extend a little beyond the outline of Mars, and we now saw that even the white light from the snow-caps had a faint tinge of red.

"For centuries the ruddy light of this planet has been remarked," said the doctor. "His very name was given him because of his gory, war-like appearance. Scientists have attempted to explain it by supposing that his vegetation is uniformly red, instead of green like ours. Still others, objecting that his vegetation could not possibly be rank or plentiful, or continue the same colour through all seasons, have supposed that his soil or primæval rock is of a deep red colour. But neither of these suppositions explain why his seas should give off a reddish light mixed with their green, or why the pure white of polar snows should be tinged with crimson."

We must have been still two hundred miles above the surface when the barometer began to rise feebly, indicating that we were already entering the Martian atmosphere; and, as we proceeded, the reddish glow spread all around us, and was even dimly visible behind as well as in front. We were still travelling too rapidly to plunge into the denser atmosphere or attempt a landing. Besides, we wished to explore the planet, and find life and civilization before choosing a landing place. And as we drew nearer, in a constantly narrowing circle, that red haze was all about us everywhere.

"There can be but one explanation of it," said the doctor at last. "This red is a colour in the Martian atmosphere. It seems very strange and almost impossible to us; but we must prepare ourselves for extremely unusual and even apparently impossible things."

But this seemed to disturb the doctor greatly, as also did the fact that we could no longer breathe with comfort the rare air which we had not found objectionable far back in space. Our returning weight made physical effort again necessary, and we were able to exert ourselves but little without panting and gasping. The rarest air we had used had shown a pressure of fourteen, and we were now compelled to increase this to eighteen in order to be comfortable.

"This Martian air is sure to give us trouble," the doctor said to me after considerable reflection. "In the first place, its red colour makes me fear it is not composed of the same gases that our air is. If it should turn out to be a mixture of oxygen and nitrogen, like ours, there is the possibility that this red matter which gives it colour will be poisonous to us. And even if it is not harmful, I do not think the air will have a pressure above ten or eleven, and we seem to need eighteen or twenty for comfort. I shall be very sorry if we have to return at once; but our supply of air is limited, you know."

"You keep a close watch through your telescope for those flying men you promised to show me," I answered. "If they can live in this air, I think we can manage it somehow. I will not go back while there is a breath left in me."

But as we drew nearer and nearer to the surface we did not discover the slightest sign of habitation. As far as we could see there was a great desert, barren of all vegetation, and apparently unwatered since creation. Our telescope did not detect the existence even of animals or creeping things.

"The wisdom of the Creator is probably quite as profound, but certainly not as apparent just here as it was somewhat farther back," I ventured.

"We must search over the whole surface of the globe until we find smoke rising," said the doctor. "That is the sure sign of intelligent life on Earth. There has hardly been a tribe of the lowest savages there which did not know how to light a fire, and this knowledge would be far more essential on a cold planet like this. Wherever we find smoke we shall find those intellectual creatures, corresponding to men on our planet."

Presently, far ahead of us, we discerned a small black cloud rapidly crossing our path. As we approached we examined it through the telescope, and soon saw that it was nothing less than an enormous flock of swiftly-flying small grey birds. This was our first acquaintance with what we afterwards found to be the predominating form of animal life on the planet. But the swift-winged cloud bore away from us, as if fleeing from the desert, and was soon lost to view.

It was not long after this that we perceived a

broad stripe of brilliant green extending down into the dull expanse of the desert. In the middle of this verdant zone there was a weaving silver ribbon, which could be nothing else than a great river, along whose banks we could discern hundreds of hovering or wading birds, hopping lugubriously, or spreading their broad wings in a low flight.

As we now lowered rapidly to examine the soil more closely, we saw that we were approaching some great geometrical masses of hewn rock, whose regularity of design indicated that they were buildings of some sort. We at once decided to land and investigate these, even if we had to take up our search for intelligent life later.

We remarked that none of these enormous structures were square, or with right-angled corners, such as we were used to. They all seemed to be a combination or multiplication of a single design, which was nothing more than a massive triangular wall, with its right angle on the ground and its acute angle at the top. Sometimes two were built together, with their perpendicular surfaces joining; again, four were joined in the same manner, and one very large one was composed of twelve of these, radiating from a common centre, which, if they had quite joined each other, would have formed a gigantic cone.

I took another look at the tall, slender birds down the river, and remarked to the doctor,—

"These great structures are no birds' nests!

You can't make me believe winged men would build with stone. These look more like giants' playthings than anything else."

"They appear to me like the gnomons of enormous sundials," remarked the doctor; "and, indeed, their uses must certainly be astronomical. With these one can not only tell the time, but the ascension and meridian of the sun and stars, and therefore the months and seasons."

We lowered and circled about above the largest one, which had twelve of the triangular walls built in circular form, with their common perpendicular line in the centre and their acute angles at the circumference. On closer observation, the twelve slanting sides, which radiated from the common peak, had a tubular appearance, and we were soon able to look down through almost a hundred great cylindrical chambers, which ran from a common opening at the top, slanting at every different angle down to the surface.

"These are nothing more than great, immovable masonry telescopes, for watching the stars in their courses!" cried the doctor. "Look, there is one perpendicular cylinder for observing just when a star or planet comes directly overhead, and these scores of other cylinders, at different angles, successively afford a view of a given constellation as it rises and then declines."

"Then they have built a separate masonry telescope, pointing in almost every conceivable direction, instead of having one movable telescope to take any direction," said I.

The wonderful size and massive construction of these was very striking, rivalling the pyramids of Egypt in their ponderous and enduring character. They were located on a raised plateau, whence the view in all directions was quite unobstructed. We came gently to land in the midst of them. To the rear, whence we had come, I could see the desolate waste of the desert. From the forward window we observed that the peaceful river kept a straight course from the cataract where it plunged over the plateau, through the green valley, between level banks, as far as we could see; and just at the foot of our plateau restfully nestled a city, whose massive and towering structures reached almost to our level. With the aid of the telescope we saw beings moving slowly about. Their form was upright and unwinged, but more than this we could not see. The deliberation and stately dignity of their movements comported perfectly with the majestic city wherein they dwelt.

"At last we have arrived at the boundaries of Martian civilization," exclaimed the doctor. "We will rest here and test the atmosphere; and if it permits us, we will then venture forth to measure our skill and knowledge against this race of builders. I hazard a guess that we will excel them in many things, for they are apparently only at the perfection of their Stone Age, while we finished that long ago, and have since passed through the Ages of Iron and of Steam, and are now at the dawn of the Era of Magnetism and Gravitation. Our minds are more fertile and elastic, for with this little movable telescope we probably obtain better results than they have done with their years of toiling calculation and patient building."

"You will be sadly disappointed if they so far excel us that they eat us up at two mouthfuls," said I. "As they move about yonder, they impress me as being full of power."

"They are as sluggish as elephants," he replied. "We are certainly more rapid in thought and action, and it is highly probable that we shall excel them in physical strength, as we have been built for three times as heavy muscular tasks as they."

"Still, if we cannot make them understand that we come peaceably as friends, they may attempt to kill us as the quickest solution of the question. And they are a whole race against two of us," said I, just beginning to realize all the difficulties that were yet ahead of us.

"Unless they are a very intelligent and magnanimous race, they will probably attempt to take us prisoners," he answered. "It is the mark of an enlightened nation to welcome strangers whose powers are unknown. A primitive race fears everything it does not understand, and force is its only argument against a superior intelligence."

Thereupon I immediately began a thorough overhauling of all the arms and ammunition, while the doctor prepared to test the air. There was a tone of confident exultation in his voice when he spoke again.

"This redness of the air will not trouble us a whit. Look! you can see no tinge of red between here and that huge wall yonder, nor anywhere along the ground as far as you can see. It is so slight a colouring that it is only noticeable in vast reaches of atmosphere, like the blue colour in our own air. See here, where a small cloud obscures the sky there is no ruddy tinge. There is no more colouring-matter in this than there is indigo in our own air. The amount of it is so infinitely small that it will never trouble us. Now, if it only contains oxygen enough, we are sure of life in it."

"Yes, if they will leave us alive to breathe it," I added, counting out seventeen cartridges for each rifle.

"The air outside shows a pressure of only eleven, while we have eighteen inside," he said. "I will bring in the discharging cylinder full of the outer air, and by keeping it upside down the lighter air will remain in it. Then, if a candle flame will burn steadily in it, the oxygen we need is there."

Suiting the action to the word, he carefully drew in the inverted cylinder, and cautiously brought a lighted candle into it. To our great delight the flame burned for a moment with a brighter, stronger light than it did in the air of the compartment.

"Hurrah!" cried the doctor, as happily as if he had just earned the right to live. "It seems to have more oxygen than our own air, which will make up for the lesser density."

Then he put the lighted candle in the cylinder, and quickly discharged it outside upon the ground where we could see it. The flame had almost twice the brilliancy that it had had inside.

"Our scientists who have sneered at the possibility of life on Mars, because of its rare atmosphere, have overlooked the simplicity of the problem. They delight in propounding posers for Omnipotence. If a Creator dilutes oxygen with three parts of nitrogen on one planet where conditions make a dense atmosphere, why should He not dilute oxygen with an equal part of nitrogen on a planet where the air is rare? Air is not a chemical compound, but a simple mixture. When a stronger, more life-giving atmosphere is needed, let there be less of the diluting gas. The nitrogen is of no known use, except to weaken the oxygen."

"Let me out into it, if you say it is all right," I cried. "I am tired of this bird-cage."

"Put on the diver's suit and helmet, and I will

weaken the pressure of the air gradually, to prevent bleeding at the nose and ears which a sudden change might cause. When you are used to the low pressure, you can throw off the helmet and try the Martian double oxygenated air."

I hurriedly donned the queer, baggy suit and the enormous helmet with the bulging glass eyes, and then connected the two long rubber tubes which sprang from the top with the air pipes which led to the doctor's compartment. He put in the bulkhead, and I went to the port-hole to unseal it. As I glanced out the little window, I thought I saw a light very near the mica. Was it our candle flame that something had lifted? The thick glass of the helmet blinded me a little, and I approached the window and peered out, coming face to face with a Martian, whose nose was pressed against the mica! What a rounded, smooth, and expressionless face! But what large, deep, luminous eyes!

I sprang back from the window in surprise, but not more quickly than he did. Just then the projectile rolled over slightly with a crunching noise, and I hear the thud of a heavy muffled blow on the doctor's end. Suddenly he pulled away the bulkhead and whispered to me excitedly:—

"They are all about us outside—dozens of them! They are examining the projectile and trying to break it open. If they strike the windows, it will be too easy."

The projectile tottered a little again. There was a heaving noise, and one end rose a little from the ground.

"They are trying to carry us off, Doctor," I cried. "You must turn in the currents and fly away from them."

The projectile was just then lifted awkwardly, and wavered a little and pitched, as if it were being carried by a throng struggling clumsily all about it. The doctor sprang to his apparatus and turned in four batteries at once. We shot up swiftly in a long curve, and from my window I could see the circle of amazed Martians, standing dumbly with their hands still held up in front of them, as they had been when the projectile left them, while they gazed open-mouthed into the sky at us.

CHAPTER II

The Terror Birds

"THEY must have thought the projectile was another chunk fallen from Phobos!" I exclaimed; "and now they can't make out why it should fly back to the satellite again."

"The more we mystify them, the more they will fear us," said the doctor. "I am going to make a swift downward swoop now, as if we would crash through the midst of them. Then perhaps they will let us alone till we are ready for them."

He had scarcely finished speaking when we shot down in a long curve, like the swing of a pendulum, apparently making directly for the group of Martians. They were not seized by any quick panic; they were too phlegmatic for that. But just as the projectile threatened to smash into them, they seemed to realize the danger, and to grasp the idea that it was being operated and directed by some power and mind inside. Then they turned, scrambling clumsily over each other, and fled with the awkward precipitation of a rhinoceros in a hurry. Our pendulum motion swung us up a little before we would have struck them, but they had scattered and were scurrying to hiding-places behind the walls of the masonry telescopes. We continued our flight to the edge of the plateau, whence we could get a better view of the city and hold a more commanding position.

"None of these who have seen our aerial evolutions are likely to trouble us again," remarked the doctor. "But they will quickly spread the news to the city, and we must be where we can watch everything that goes on there, and hurriedly prepare for the worst they can do to us. We will seek the principal approach to the plateau and defend it."

His ideas had suddenly become altogether war-like. I liked the excitement of it so far, and hastened to agree with him. We came to land in a sheltered part of the main road leading to the plateau, and prepared to emerge and set up our telescope where it would sweep the city.

"Shall we try this air on the dog before you go out?" inquired the doctor in all seriousness.

"Try it on the rabbit if you wish, but not on Two-spot."

He put Bunny into the discharging cylinder and pushed him out. The meek little animal seemed quite delighted at being released. He hopped about playfully, skipping much higher and farther at each hop than I had ever seen him do before.

This reassured me, and I put on the helmet

again, and opened the port-hole. As the rarer Martian air swept in, my suit swelled and puffed to its fullest capacity, by the expansion of the denser air within it. I was so blown up that I could scarcely squeeze myself out of the port-hole. It was like a red misty day outside, though there were no clouds. The sky was a perfectly cloudless dull red, and the coppery sun was shining almost overhead. His orb looked less than two-thirds the size it did from the Earth, and one could look at its duller light fixedly without hurting the eyes. Phobos was also faintly visible, steering his backward course across the ruddy sky. The thermometer showed a temperature just above freezing, but I was perfectly warm within the diver's suit and its envelope of air. The red haze and utter lack of breeze added a deceptive appearance of sultry heat.

I was gazing back toward the Gnomons, when suddenly a group of the Martians we had first seen came around a turn of the road and over a knoll into full view of us. They were plainly surprised beyond all measure by my strange appearance. My puffed and corpulent figure, my bulging face of glass, my two long rubber tentacles extending back into my shell, must have made them think I was a very curious animal! Also they were probably surprised at seeing any living thing come out of the mass, which they must have thought had fallen from their moon, for she was always shying things at them. And I now had my first chance to study their appearance closely.

"Doctor," I said softly, to see if he could hear me through the connecting tubes. As I had hoped, they proved to be very good speaking-trumpets, and I heard his answer noisily.

"Speak lower; I hear you easily," I said. "There is a party of them coming down this road to descend to the city. They have stopped upon seeing me. They are nothing but men like ourselves. I see no wings, horns, tails, or other appendages that we have not. They are just fat, puffy, sluggish men, very white and pale in colour, and covered with a peculiar clothing that looks like feathers. I seem to be a far greater freak to them than they are to me."

Had he been a million miles away, I should have known that it was the doctor answering, from his unsurprised and matter-of-fact tone. I imagined I could see the exact expression of his face as he said,—

"After all, then, man is the most perfect animal the Creator could make. From a mechanical standpoint he needs nothing that he has not, and has nothing that he does not need. However you change him, you would make him imperfect. Physiologically he may be much the same on all the planets, but there is room for the widest variations on the intellectual and spiritual side."

"Do not forget that my patriarchal ancestors

record that God made man in His own image, upon which there could be no improvement," I put in.

"Yes, but modern scientists would have us believe that your patriarchs would have written a different fable if they had understood the theory of evolution. It appears that man is really a little lower than the angels, by being material and ponderable and visible, but the general image may be the same. Perhaps upon the various planets it may be that the same lines of differences prevail, as between the heathen tribes and the civilized people on earth. There at least we are sure that physiologically no marked difference exists between the lowest savage and the wisest sage."

"Except, perhaps, that the savage may have the best digestion," I added. "These look as if they had but few troubles and plenty to eat. I see no wrinkles or hard lines. Their forms and features are gracefully rounded. Their eyes are larger and stronger, with a liquid depth suited to this soft and weaker light. None of them wear beards, and very little hair is visible. I must say they do not look at all warlike. If we could only make them understand that we are friendly, I think they would gladly bid us to a feast of freshly-cooked meats and good wines, and ask us, chuckling, for the latest after-dinner stories that are current on Earth."

"Make friendly signs to them, and see how they behave," he suggested.

I slowly waved my hand to them to approach, and extended my arm as if to shake hands. While talking with the doctor I had stood perfectly still, and they had been warily watching me all the time. When I moved and stretched out my arm, they took fright and fled precipitately.

"I have scared them away, as if they were a lot of roe deer!" I exclaimed.

"Then let us hasten preparations while they are gone," he replied. "If you can stand the pressure I have given you, it will be safe to throw off the helmet and suit."

Upon lifting the cover from my head, I caught a draught of fresh cold air that was unspeakably invigorating. I drank it in deep breaths, and felt like skipping about for joy. Kicking off the suit that trammelled me, I put it and the helmet back inside and closed the port-hole. Then the doctor pulled away the bulkhead and breathed the mixed atmosphere, half-Martian from my compartment and half-Earthly from his. He suffered no inconvenience from the sudden half-way step toward a lower density, and presently he emerged into the exhilarating air with me.

"This atmosphere has a stimulation in it like thin wine, and it gives me an appetite. I feel strong and virile enough to tip Mars topsy-turvy," I said. "At least, let me get some cigars to smoke while we are arming our stronghold."

When I went in for the guns, I put a handful of

Havanas in my vest pocket, and emerging, I laid the rifles handy and proceeded to light a weed. I was watching the bright flame of the match, and puffing with gusto at the fragrant smoke, when from another direction a second squad of Martians came into view very near us. They immediately halted and gazed at us in open-mouthed wonder, which soon changed to a look of horror. Remembering the pipe of peace among the American Indians, I drew out a cigar, and hastily striking a match upon my trousers, I held the weed and flame toward them. Not a man of them stayed to see any more. Their flight was more precipitate than the other party's had been.

"It was your smoke they were afraid of," said the doctor. "Whenever you puffed, I saw them looking at each other blankly and dropping back a little. They have taken you for a fire-eater and a smoke-breather, and when you drew the flame from your lungs it was too much for them. But all this serves our purpose of frightening them. They will spread strange stories in the city below!"

I helped him carry out the telescope, and we placed it in a commanding position. Then we propped up the broad shields, so that each of us could crouch behind one, and I laid a broadsword and rifle handy to each. Then we put on the linked-wire shirts under our coats, buckled the revolvers about us, and, as it was rather cold, we each put on a thick pair of gloves and a heavy topcoat.

The doctor, who was carefully watching things down in the city through the telescope, cried out to me presently,—

"There is wild commotion and great excitement down yonder by the great palace. The news has reached them! They are preparing to come in force to take us!"

"I wish I knew what their sign of peace is, we might save a conflict," said I. "Perhaps our firearms won't harm them."

"More likely they will blow them all to pieces," answered the doctor. "But we must not fire unless it becomes absolutely necessary to defend ourselves, for if we kill any of them, they will then have cause to deal with us as dreadfully as they can. We cannot hope to overcome them all. It will be enough to demonstrate our supremacy, so that they will allow us to live among them. Therefore, let us simply defend ourselves and do nothing offensive, thus showing that we are peaceably disposed."

"You cry peace, but look at the great army they are sending against us!" I exclaimed. "There are four companies of foot soldiers marching through the streets, and each man is armed with a very long cross-bow and wears a brightly-coloured birdwing on his forehead. The streets are filling with people to see them pass. Now three more companies wheel out of the palace, but they have no cross-bows. They are whirling something around their heads."

The doctor anxiously awaited his turn at the telescope, and as he looked he clutched his pistol though they were still several miles away.

"Those are slings they are whirling about their heads," he said. "And the commander of each company rides an ambling donkey, and wears a heavy plaited beard and long braided hair, without head covering."

"But look further back, coming out of the palace now!" I cried. "What are those strange, stately animals far behind the soldiers? I can see them with the naked eye."

"*Donnerwetter!* what towering birds!" he muttered under his breath. "Like ostriches in form, but as tall and graceful as a giraffe! There is a man riding astride the neck of each of them, yet he could scarcely reach half-way to their heads!"

"Are those monstrous things birds?" I demanded. "Let me look. What long and bony legs they have! They would stride over us without touching our heads; but how they could kick!"

"And how they could run!" put in the doctor. "See, they stride easily over seven or eight feet with a single step. They must be messenger birds, for there are only four of them, and their riders are not armed."

"They may have hundreds more of them in reserve, and they could fight far more viciously than the men. See what a wicked beak and what a long muscular neck they have. They could crush a skull in a twinkling with one swift swoop of that head! I will fight the men, but I will take no chances with those birds!"

Although these strange, small-winged creatures had started long after the soldiers, they had quickly passed them, and were now beginning to mount toward our plateau. They were making swift detours at intervals, as if to reconnoitre. We were hidden behind our rocks and shields, and the riders could not see us, and they had evidently not yet seen the brass barrel of our telescope. It would be folly for them to attempt to come up the road we were guarding, for we could easily heave boulders over and crush them. I had already put my shoulder to an immense rock near the brink, to see if it was as heavy as it looked. I found it porous and crumbly, and no heavier than so much chalk. Up the roadway the great birds climbed with wonderful ease. Their riders were evidently looking for us without any idea where we were.

"I won't see those elephantine bipeds come any nearer to me!" I exclaimed, and rushing to the boulder, which was certainly four feet in diameter, I toppled it over the brink, and expected to see it carry everything down before it. It rolled slowly down the steep bank, with hardly a third the force and speed of the same mass on Earth. This discouraged me, but I watched for it to reach the foremost bird. He was surprised by it, but made one step sideways, and, lifting his great right leg, the stone rolled under him without any damage. He gave a queer, guttural croak, accompanied by a most violent motion of the head and neck. The other birds, thus warned, dodged quickly sidewise, and avoided the slowly rolling boulder; but all three of the riders were thrown by the swift lateral movement of the birds. The astonished men picked themselves up slowly from the bushes and approached their birds. But they could scarcely reach with their hands the lower part of the neck where they had sat.

"Unless they are good jumpers, they cannot mount again without a ladder!" said the doctor.

"Jumping is easier than standing still here," I interrupted. "I can jump ten feet high with no trouble."

"Yes; but these Martian boobies haven't your muscles. *Aber Blitzen!* did you see that fellow mount his bird again?"

I had seen it, and I do not remember anything more wonderful than this operation, which was repeated for each rider. The man went in front of his bird, turned his back, and stooped forward. The bird then curved his long neck to the ground, and put his head and neck between the legs of the rider, who clutched tightly with his arms and legs.

With a swift, graceful swing, the bird lifted its head on high, carrying the rider as if he were nothing. When the great neck was again erect, the man slid carefully down it to his place, much as one might slip down a telegraph pole. Then two of the birds turned back to the city as swiftly as they could go, and the other two took separate side trails and soon disappeared.

CHAPTER III

The Armies of Mars

AS the two returning birds passed the marching soldiers, their riders evidently delivered some message to the captains, for the soldiers suddenly broke forward in a run, using their long cross-bows with great dexterity as jumping staves. Placing the outer end upon the ground ahead of them as they ran, they leaped and hung upon the cross-piece with their hands. The springy resistance of this tough wood imparted to them a forward motion with its rebound, and they scaled great distances at each jump. The whole company did it in concert, and they made almost as great speed as if they had been riding bicycles. The slingers were consequently left far in the rear.

Less than half way up the incline the archers stopped, arranged their bow-thongs, and selected feathered arrows from a pouch slung over their shoulders.

"They can never hit us from that distance!" I exclaimed; "a rifle would not carry so far."

"You forget the weak gravity which will bend

their course down very little, and the thin air which will barely resist their flight; this is a model planet for archery," he answered. "Quick! drop behind your shield! They have fired the first volley!"

A torrent of the shafts fell all about us, and many pelted against our shields. Those which struck the soft earth of the bank sank into it and stuck there, but those which struck our steel were shivered and broken.

"Sit still and let them shoot away their arrows," I whispered. "This will soon be over."

The next volley came with a little more force, as if they had marched further up the hill. One or two arrows fell very near me, and I reached for them to examine their construction. They were made of the hollow, filmy stock of a rather tough reed, and were pointed with a chipped stone tip, which was brittle, but not harder than porous chalk.

"That stuff wouldn't pierce my two coats, to say nothing of the linked steel shirt," I sneered. "I will show them what fools they are!" and I walked boldly out to the brink and faced them. They let fly a quick volley with a concerted shout. As I saw the arrows start, I turned my back and bent down my head quickly. Perhaps a dozen of the slim reeds pelted me, and then I stooped over and gathered up as many as I could find, and broke them all in my hands before their eyes.

This sent a hum of excited jabbering through their ranks, and they fired no more. I stood watching them, and presently I grasped my two hands together and shook hands with myself, to try to convey to them the idea that we were friendly; but it must have carried no meaning to them. By this time the slingers had come up, and I retired behind my shield to await their action. The archers seemed very glad of their arrival, and yielded the foremost place to them. I noted their operations carefully, and saw them place something, which did not look like a round stone, in the pocket of their slings, and then they whirled it long and cautiously. Suddenly they discharged it with a swift movement of their bodies backward, which landed them on one knee.

"Wide of the mark!" I cried, as the missiles sailed off far to the right of us. But just before landing they bent a sharp, surprising curve, and lacked but little of hitting us behind the shields! The things they had thrown were the thin, concave shells of a large nut, and the trick of discharging them gave them their peculiar flight.

"I don't like this throwing around the corner!" exclaimed the doctor. "With a little truer aim they will be able to hit us behind anything."

"Hurry, bring your shield over behind mine, and face it the other way," said I; "then we will crouch between the two in safety."

He did this just in time, for some of the next

volley actually curved around and hit his shield, but none struck mine in front. However, the shells which fell near us were of light weight, and would not have bruised us much with heavy clothing on. Presently their pelting ceased, and we concluded that they were planning something new. We decided to let them know that we were not hurt, so we emerged; and I tried throwing the shells back with my hand, but I could not control their erratic course. When they saw this they jeered at me, and I itched to treat them to just one pistol shot, only to show them what child's play their fighting was! Presently we saw what they were waiting for. Far down the road the two great birds were returning harnessed together, and dragging behind them an enormous catapult. Tied across their backs were two stout darts, seemingly twelve feet long and three inches square. Each of them had a wicked-looking barbed tip.

There was a pleased and confident jabber among the slingers and archers below as the birds arrived. The catapult was turned about toward us, and lashed tightly to stakes driven in front and behind. Then the birds were hitched to the cord of the immense bow, and they pulled it far back, until the men made it fast in a notch. The cross-piece had now become almost a half-circle, quite ten feet in diameter. The captain of a company of archers acted as gunner, and carefully adjusted the catapult, aiming it evidently at our shield. Upon seeing this we placed the two shields together, and leaned them both inward toward us, so as to make their angle with the upward course of the dart more obtuse, and thus cause a glancing blow instead of a solid impact. Crouching under the steel shelters, we awaited the dart.

Whiz-z-z it whistled up through the thin air! Bimm-m! it struck the top of our outer shield, and glanced off as we had hoped. The outer steel rattled and banged against the inner, and both shields pressed hard over against us, but not the slightest damage was done.

We went out to watch them load the second dart. They evidently saw the impotence of the glancing blow, and were noisily discussing it. A captain of the slingers was arguing hotly with the gunner, who was finally persuaded to take his aim a little lower. Then a hum of approval went through the throng.

"They do think a little, but they are not secretive!" I sneered, flopping our inner shield over flat on the ground. "Come, sit on this, Doctor, and we will lean the outer shield over us, and snuggle in between them as cosy as two oysters! Let them fondly imagine they can shoot us through this pasty soil, and keep their own counsel better after this!"

It was not a bad guess on my part; for the second dart struck the edge of the cliff, bored through the loose soil, and thumped our lower shield with a dull thud that lifted us from the ground. But the point and edges of the dart were blunted, and crumbled with the blow, and I could find no dent in the shield.

"See, the birds are returning to the city in haste for more darts!" said the doctor. But I was interested in examining the first dart, which had fallen a few hundred feet behind us. Its shaft was of roughly-hewn, spongy wood, and it weighed far less than half the mass of soft pine would on Earth. Its tip was not metal, but chipped stone—crumbly, like the arrow-heads. Either they did not know the metals, or they were too rare to be used in their arts. And it was to be supposed that they would use the hardest stone they had for arrow-heads and dart-tips.

I carried the shaft easily upon my shoulder forward to the edge of the cliff. This surprised even the doctor a little, for four Martians had been necessary to put it in place upon the catapult. It must have astonished them still more, for they were staring at me so blankly that I was tempted to toss the dart down their gaping throats!

"Give them just one dose of their own medicine!" suggested the doctor.

"Perhaps I had better teach them to keep their dangerous weapons at home," I said; and, balancing the dart easily above my head, I aimed it carefully at a dense group around the catapult. I threw my whole force into the thrust, and sent the shaft whizzing down at them. Then I staggered back, quite exhausted by the effort and gasping for breath.

"Good God! You have impaled two of them upon the dart!" cried the doctor, "and it is causing a panic in the whole army!"

And when I sprang up to look, I saw two writhing Martians, much shrunken in size and dying upon the dart. The terror-stricken archers and slingers were scattering and scurrying in every direction, regardless of the shouted orders of their captains. The foremost of the impaled men wore a beard, and was no other than the gunner of the catapult.

"I am sorry for the poor devils!" I exclaimed. "I had no idea they were so soft and tender. They have shrunk like a pricked balloon!"

"They thought they could prick us like that, and let the life ooze out," said the doctor. "There is no danger that they will shoot any more at us. The whole army is afraid that you will throw down the other dart."

Nevertheless, other companies of archers and slingers were seen leaving the palace, and the birds were already returning with two more darts. And the soldiers below were gaining courage and responding to the rallying cries of the captains, who were halloing and pointing toward the edge of the cliff, down in the direction of the cataract. I looked quickly that way, and instantly shouted,—

"To the rifles, quick, doctor! The other two birds have ascended the cliff, and are racing toward us along its edge. Take careful aim at the head of that front one. Afterward, let drive two random bullets into his body!"

Urged on by their riders, who with their hands swayed the long necks of the birds in unison with their rhythmical stride, these two-legged giraffes, with the wild look and sharp beak of an eagle, swept menacingly toward us.

"Ready now!" I cried, as the foremost came within fifty feet of us. "Fire!"

Two sharp reports almost simultaneous, with a less thunderous explosion than on Earth, but singing in a higher key and flaming vastly more, startled and terrified the Martians. Then crack! crack! bang! bang! four other shots in swift succession, followed by the terrific croaking of the wounded Terror-bird, which fell ponderously forward, kicking violently and beating the ground wildly with its head.

Seizing my broadsword in a flash, I dealt it such a blow upon the neck as quite to sever the head from the body. There was a gush of red blood; and those who have seen the antics of a decapitated chicken, may correspondingly multiply the corpse and imagine the confusion that now ensued.

"Stand ready for the second bird!" I shouted to the doctor; but on looking, I saw that the other animal refused to be urged forward, after seeing the fate of his companion. His rider was half-hearted in his efforts, and was watching the forward rider, who had been severely thrown with the bird's fall, and badly bruised by the kicking and threshing. He seemed to realize that he was in our power, and was thoroughly desperate. With a wailing cry he rushed at me with open arms, as if to embrace death, for I still held the sword. Dropping the weapon, I grappled with him, catching him about the wrists, which shrank under my grasp. He seemed to have scarcely the strength of a child; and everywhere I touched him, his flesh yielded like the flabby muscles of a fat baby. I bent him over backwards, then swung him around and caught him by the shoulders, and whirled him around my head. Finally, I tossed him over the edge of the cliff, where he landed among some bushes, and scrambled down as fast as he could, glad to have saved his life. The other rider had turned his bird back toward the cataract with all possible despatch.

"The whole army below us is now thoroughly demoralized!" said the jubilant doctor. "Many of them fled dismayed on hearing the firing, and others screamed and ran away when they saw you decapitate the bird. But your wrestling with the rider, and flinging him about like an infant, was an object lesson none of them could stay to see repeated. I saw one trembling fool slink back to cut the thong of the catapult, so that we could not use it on them They have wholly abandoned the attack!"

"If this is the worst they can do, I will undertake to make myself king, and you prime minister here, within twenty-four hours!" I ejaculated, decidedly pleased with the idea. "And I will maintain supremacy with a standing army of a thousand Terror-birds!"

"The consciousness of superior strength always brings that desire for conquest," answered the doctor. "We must not allow it to master us, but we must push our advantage. Look! the panic of the first ones reaching the city is spreading to the new companies marching out. They are trampled over by the fleeing host, they turn and mingle with the frightened mob in one struggling, terror-stricken mass! Come, let us be into the projectile and after them. With a few booming shots above their heads, we will make them think their Thunder-gods have come!"

CHAPTER IV

The Strange Bravery of Miss Blank

TELESCOPE, rifles, and shields were tumbled into the projectile pell-mell, and without stopping to close the port-hole, we steered towards the city as we mounted rapidly. When the soldiers, weary of running, saw us start, they were stricken with a new fear, and made all possible haste for shelter. When they perceived that we were rising into the red haze, they took a little courage, but still hastened.

"Perhaps they think we are mounting to the sky for more thunder and lightning," I suggested. "Little do they know the destruction we could do them with the handful of ammunition we have, if we really meant war as much as they at first desired it and now fear it!"

By this time we were almost above the thickest crowd of the fleeing army, while the most energetic runners and the Terror-bird that had turned back had reached the heart of the city; and we could see the alarm spreading like wild-fire to all its inhabitants. I was busy loading the rifles with the cartridges which the doctor had robbed of their bullets for the pickle-bottle experiment soon after our start.

"We will execute a little *coup*, to show them the difficulties of retreat when the enemy is armed with gravity projectiles," said the doctor. "Do you see that great gate of the city they are all making for? We will drop down there, just in front of them, and prevent their entrance. It will be better to keep the whole army outside the walls, if possible, for its absence and disorganization will make the rulers all the more tractable when we are ready to drop down into their city and make peace with them on our own terms."

"I must say you are a good general, Doctor!" I exclaimed. "You plan the campaign, and I will do the fighting."

The blank dismay of the soldiers when they saw us descending again, and their abject desperation when they perceived that we should land in front of them and cut off their entrance to the city, was pitiful to see.

"Doctor, do you remember the grand display and the proud strength with which these soldiers marched forth? Look at the difference now!"

"Oh, war! war!" he exclaimed. "The glory of its beginning! The terror of its prosecution! The misery of its end! Would that it could always be carried on by terrorizing the mind instead of by slaying the body!"

As we were about to come to land in front of the struggling multitude of soldiers, I fired a dozen blank cartridges as rapidly as I could work the rifle. This was at very near range, and although the explosions sounded weak to me, the excessive flaming of the powder added a new terror. The disorganized army stopped in dread; the stragglers pushing up from behind, and the frightened turning of those in front, crushed the multitude together and increased the confusion. Throngs of people, whose curiosity was still stronger than their fear, were coming out from the city. As they saw us float down and land, and then heard the firing, they turned and rushed within the gates again, ready to believe far worse stories than they had yet heard.

"We must scatter this rabble army and put it wholly to rout," insisted the doctor. "I will swing amongst them and over their heads, while you burn powder for them. If they won't scatter, use your revolver and wound one or two of them."

"No, I will not harm another man," I answered. "They are too weak and defenceless a foe, and are no match for us. Hereafter I will fight only with the birds."

We rose and sailed slantingly toward them, but they had already started to disperse. Those who had jumping-staves disentangled themselves from the crowd and scattered into the bushy wastes. I continued firing until my blank cartridges were gone, and then we landed just outside the entrance and emerged from the projectile to examine the gates and see if we could close and fasten them.

Within the wall those who had gained entrance during our last movements were rapidly retreating toward the centre of the city, warning all whom they passed. One single stately figure showed no fear, and paid no heed to the exclamations of the runners. The ampler dress and flowing flaxen hair indicated that it was a woman, and to our surprise, though she was well clothed, she seemed to be demanding alms of every one as she approached us. No one gave her anything, and occasionally a runner seized her arm and tried to persuade her to return. But she caught none of their excitement, and composedly pursued her course.

"Egad! This beautiful girl is braver than the whole Martian army!" I exclaimed in amazement, as she calmly approached where I was standing by the gate and extended her fair, plump hand. If she was asking alms, I had nothing to give her; but here, at least, was one pacific, composed, and reasonable person. Perhaps it was the queen, or a diplomatic envoy of the ruler!

"Now, is the time to demonstrate our friendliness," I exclaimed, and reaching forth my hand I grasped hers in a warm clasp of welcome.

She looked up at me blankly. Her beautiful face carried no expression of satisfaction or surprise. Her transparent complexion was neither paled by fear nor flushed by pleasure. Her great dreamy eyes, of a deep liquid blue, wandered unfixedly in their languid gaze. Still holding her soft hand, which was far warmer than my own, I opened her fingers with my other hand and pointed at her pink extended palm as if to inquire what she wished. I watched her closely, but she made no sign, said nothing, looked nothing.

"Since I do not know you, I can think of no more fitting name to call you by than Miss Blank," I said, more to express my thought in articulate sounds than anything else, for I had no idea she would understand me. From her expression I could not judge whether she had even heard me, to say nothing of comprehending. She was looking beyond me, through the gate, as if searching others from whom she might ask alms. Seeing none, she wheeled slowly about to return. Unwillingly I released her hand, and stood unspeakably puzzled by the whole matter. She was commanding in appearance, being taller than I by a few inches, not slim, but well proportioned. She had the stately serenity of a dreaming queen, but the blank, unresponsive soul of one who dwelt within herself; and though she saw, she did not realize the existence or meaning of anything outside.

"Doctor, will all your learning solve this riddle for me!" I exclaimed. "Can all the Martian women be like this? She is beautiful of body and strangely warm and winning to the touch, but as cold of heart as the drifting snow that suffocates a poor lost lamb. She has had a strange influence over me; a puzzling, baffling attraction. A suggestion of something delicate and subtlely charming, which, when one seeks to seize and to define, retires icily behind the drawn curtain of her soul."

"I hope you won't play the lost lamb to her snowdrift!" he sneered, in a way that I resented. "One would think she had hypnotized you on the spot! And she must be in a trance herself, for she had not sense enough to fear us."

"Those who have the most sense fear us the least!" I retorted.

"But fear is our sharp weapon now," he answered; "and some of the stragglers, looking back, saw you stand there holding her hand in a manner far from warlike. They will report this to the rulers unless we forestall them. Come, fasten the gates tightly upon the inside to keep the soldiers out, and I will sail over the wall to pick you up."

"Doctor, we make our peace at once, and fight no more with the brothers of this girl," I said with decision.

The massive gates were of hewn stone, turning in sockets at their outer corners above and below. They swung as easily as if hung upon hinges, and when closed a slab of stone came down to bar them. I made them fast, and then called out to the doctor,—

"Don't come for me. I have found a jumping-staff, and I think I can leap to the top of the wall."

It was a sheer fifteen feet of solid masonry, but my chief delight since landing on Martian soil was the inordinate springiness of my leg muscles against the feeble gravity. I ran and sprang lustily with the aid of the cross-bow, and I remember the doctor's surprised look when he saw me clear the entire wall without touching the top and land safely with a very mild jolt on his side.

A short oblique ascent of the projectile brought us over the city, and revealed to us the condition of desperate panic into which the wild reports of the soldiers and the bird-rider had thrown the frantic populace. The soldiers still within the walls could not restrain the people, or did not try. If there was any government, it lacked a head or could not command attention. The stubborn instinct of self-preservation was king. Distracted throngs surged out at one gate, to separate and waver and hesitate, and finally to fight for a speedy entrance at another. On one side soldiers were apparently ordering people down from the wall, while on another the excited populace was hauling sentinel soldiers from the same elevation, lest our attention should be attracted. Within, strong men were weeping and wailing; without, nervous men were haranguing the vacillating multitude; but more were stolidly pushing with the rabble or being hustled by it.

Only one sign of order and forethought was apparent. Evidently for better safety and for an easier defence, the women and children had been taken to a central park or pleasure ground, and left there with a small guard of soldiers. The men to whom they belonged had apparently all gone elsewhere.

"Doctor, we must put an end to this fear and frenzy at the earliest possible moment. If we are not destroying those people, we are exciting them to destroy each other, which is equally blameworthy. We must go down at once, but we had best avoid the frantic men. The women seem far more reposeful. Let us drop quietly into that open field in the park, and I will make friendly signs to the women, pat the children on the head, and give them all to understand that we mean no harm."

He evidently saw that we had quite overdone the scare, and was as much impressed by the terrible picture below as I was. We turned down without delay, and landed quietly behind a clump of trees. I took a tin of sweet biscuits under my arm, and the doctor following me, with a generous handful of his trinkets and tinsel toys, we left the projectile, and rounding the grove of dwarfed trees we approached the romping children first. I patted their flaxen curls, lightly pinched their cheeks, and handed each of them a sweet biscuit. Then, while the doctor distributed strange toys amongst them, I put on my most courtly ways and addressed myself to the women. Their first impulses of fear had been somewhat allayed by our attentions to the children, and I bowed profusely and made bold to kiss the hands of a few of the youngest of them. Each of these looked to see if I had left anything visible or harmful on her hand, from which I judged the custom was wholly strange to them. The others looked on askance and whispered excitedly among themselves.

One of the soldiers who had seen us approach, but offered no resistance, had now started to run, as fast as his jumping-staff would carry him, toward the palace. I knew at once that this meant some new development, and I hoped it meant a report of our friendly actions and a truce all around. But the doctor reminded me that we must be prepared for surprises and treachery. Therefore we re-entered the projectile, and out of the sight of all the Martians I re-loaded the rifles, and then we waited a long time.

Our patience was finally rewarded, for we saw the soldier returning, slowly leading a woman. In her left arm, which the soldier held, she carried something white which wriggled occasionally. All this we considered so favourable a development that we went out again, bowing to the women about us, petting the children, and looking as peaceable and amiable as the politest of Earth's people. But it may have passed for imbecility, or worse, on Mars.

When I looked toward the soldier again, my heart began a queer thumping, for he was leading no other than the woman who had met us at the gate, and she was carrying our white rabbit, which we had released early that morning a long way from this spot.

"By all that is wonderful!" I exclaimed to the doctor, "if we have not fallen upon a country which is ruled by yon dumb queen, and she brings to us as a peace offering the only thing that we have lost!"

"Since when have potentates learned to beg, and forgotten to command and to exact?" he answered with half a sneer. "See, she still extends her hand to every one she passes."

And as the soldier, trained to revere a beard, led the woman directly up to the doctor, she stretched forth her pretty palm again; but if he had presumed to take it I could have struck him! To my cordial grasp I added a kiss this time, and then I raised my eyes slowly to her face, fearing to see that blank look again. There *was* no look in her eyes; they did not look, they only wandered!

The soldier, who still held her other arm, waved his cross-bow toward the palace meaningly, and a hush fell upon the murmuring crowd. I ignored him and spoke to her,—

"If thou art the queen, command me but by a look or sign, and I obey. And if thou art not the queen, then they should make thee one. Dost thou wish us to follow thee to yon palace?" said I; but the only mind that understood scoffed at my rapturous declamation.

The woman merely drew her hand from my warm clasp and stretched it out to the people, who crowded about and paid her no attention. Then the soldier, as if suddenly remembering, took the rabbit from her arm and handed it to me. She looked about at this, as if missing the snuggling animal, and I stared hard at the meddling soldier to reprove him for interfering with his queen, and gently restored the rabbit to her arm.

"The soldier wishes us to go to the palace," put in the doctor. "But we must not go unarmed. He may be leading us into an ambush. Let us take all of our arms and follow him."

Accordingly, we buckled on the swords, and took the rifles on our shoulders. As we dragged out the heavy shields, the soldier pointed to a group of donkeys laden with bags of something like grain. I waved assent, and the muleteer unburdened one of them and loaded the shields upon him.

"Why not take the telescope?" I suggested; "it is big and bright, and perhaps they may fear it too. Or we may wish to show its wondrous use."

As I drew it out the crowd started back, but the soldier and the muleteer gingerly loaded it upon another donkey. Then the soldier took the woman's arm again, and pushed her extended palm around toward me, as if I would be unwilling to go unless I had it. My right hand held my rifle, but I was secretly glad that my left was free to clasp the woman's hand. The doctor walked behind to watch the muleteer, and thus we marched to the palace.

CHAPTER V

Zaphnath, Ruler of the Kemi

TWO hieroglyph-bearing columns of red sandstone, strong and broad enough to have supported a Tower of Babel, formed the portals of the outer gate of the palace. A pair of Terror-birds, whose plumage was a pearly grey, stood sleepily on guard. Our soldier, who could scarcely have reached to the backs of the birds, lifted up his cross-bow and tapped upon their long necks. Acting perfectly in concert, the animals each engaged with its beak a wooden ring suspended high in front of them, and then, bending down their necks, the hempen ropes, to which the rings were fastened, hauled up a ponderous portcullis, made of slabs of stone, and thus afforded us an entrance.

As this stone gate rumbled slowly down again, we saw that we were shut into a vast courtyard, surrounded by a colonnade, whence cavernous passages led circuitously to the various compartments of the palace. Within the courtyard were drawn up in expectant readiness four companies of archers and three of slingers, in all, perhaps, seven hundred men, who gaped and stared at us.

The doctor touched my elbow, and whispered: "We should have landed in here with the projectile, which would have given us a means of ready escape."

"Remember the saying of General Grant," I answered. "'When you are frightened, don't forget that the enemy may be far more so.' These soldiers have heard enough to make them believe us capable of anything. They would tear down the very walls, if we were to open fire on them. Besides, I could leap that courtyard wall and drag you with me."

Unsheathing our swords, as an object lesson to the soldiers, we followed our guide to the blind end of a long passage, which apparently gave entrance only to a small stone chamber. Following the soldier and muleteer, who were now carrying our shields and telescope, we crowded into this and waited. Presently the entire chamber, operated in some unseen manner, turned slowly half way round, so that its door now gave entrance directly to a vast but gloomy and tomb-like audience chamber, where we were evidently expected.

Upon a massive throne of richly-chiselled stone a youth of scarcely more than five-and-twenty years (if judged by earthly standards) sat gorgeously arrayed in vestments of richly coloured feathers, woven skilfully into the meshes of coarse cloth.

Longer plumes of changeable colours radiated from a wide collar which he wore, covering his breast and back, and extending over his shoulders. The peach-blow of his fair cheeks was partly hidden by a heavy false beard, plaited into stubby braids, which hung to an even line a little below the chin. His own soft, flaxen hair peeped meekly out from under a wig of tightly curled grey strands, cropped all round to a level with the beard. His feet and arms were bare, except for thin ribbons of downy, purple feathers, which circled the wrists and ankles. No crown was on his head, but among the stringy wig-curls the sinuous body of an asp bent in and out, and the curved neck and threatening head surmounted his clear brow.

To his right, round an oval table of highly polished stone, sat twelve wrinkled men, not one of whom but had seen three times his years. They wore their own white beards, unplaited, and their feather clothing was less elaborate and of simple grey, like the plumage of the Terror-bird.

Our soldier placed his right hand upon his cheek, and inclined his head slightly forward and to the right, as a salutation to the ruler, and, leaving the woman standing by me, he and the muleteer retired. She seemed neither surprised at, nor accustomed to, these surroundings. She made no salutation or obeisance to the ruler or to the old men, and they made none to her. Withdrawing her hand from mine, she stretched it toward them, as she had toward the commonest man outside. They paid her no attention, but the oldest of the men signalled to an attendant, who led her back and placed her hand in mine again. That soldiers and counsellors alike should consider this necessary or fitting seemed strange to me. The doctor jokingly suggested that they wished to keep me permanently hypnotized, lest I should become dangerous again.

Having laid off our rifles, swords, and outer coats, I lifted my cap and made a low bow to the youth and to the old men, but the doctor tried the salute of the right hand upon the cheek, as he had seen the soldier do. In answer the youth simply looked toward the twelve, waving his hand towards us in a way which seemed to say to them, "Gentlemen, behold the enigma!" Then, beginning with the eldest, the twelve jabbered at us in turn, apparently in different tongues, some sibilant, some guttural, and others with the musical cadence of frequent vowel sounds. Needless to say, each was equally incomprehensible to us, and we did not think it worth while to try German or English upon them. When they had finished, they looked much vexed, and slowly wagged their beards. Then the youth spoke something to them with a confident gesture toward himself. He arose, and began addressing us. I suddenly stopped short in the middle of a sentence I was whispering to the doctor. It seemed as if the youth had ceased making mere sounds, and had begun to speak a coherent language, a tongue which has lived ages while others have languished into forgetfulness; a language whose words I understood, but yet the words carried little clear meaning to me.

"Listen, Doctor! The boy is speaking Hebrew! Ancient and archaic in form, but yet Hebrew which I understand!" And this is what he had said:

"Oh ye, who speak among yourselves, but understand only those who speak not at all, I, Zaphnath, revealer of God's hidden things, will address ye in my native tongue, which none but me in all the land of Kem hath any knowledge of."

"There be two of us in Kem, O Zaphnath, who understand that tongue. Speak on!" I cried.

But the boy stripped off his wig and beard, and, leaving the throne, hastened toward me and laid his soft right cheek against my own with gentle pressure.

"Comest thou, then, from the land of my father, a stranger wandering into Kem, even as I came?" he asked.

"Nay, gentle youth, we came a vastly farther way, from another world, so distant that thou seest it from here only as a twinkling star in the night. But if, indeed, thou camest a wandering stranger into Kem, art thou then the king?" He had resumed his wig and beard, and his proud seat upon the throne, and after he had translated my words for the twelve old men, he answered me,—

"I am Zaphnath, ruler over all the land of Kem,

without whom the Pharaoh doeth not, nor sayeth anything. These are his twelve wise men, who do not believe what thou hast said, for there is no other world large enough for the abode of two men, except the Day-Giver, whence they think ye have come. The Pharaoh may believe them, but I will believe what ye tell me. He hath given me full power to treat with you, and hath taken refuge with all his women in his tomb, and will not come forth until ye be appeased. Tell me in truth, then, are ye men, or gods? Ye look not half so warlike as all the soldiers have described you."

I translated this to the doctor, but replied without waiting to consult with him,—

"We know but one God, who hath made all the stars, and all who dwell upon them. We are men to whom it hath been given to travel the infinite distances which reach from one of His stars to another, and we are come to this one, not to make war but to find peace. We would have sought thee peacefully as friends, had not thine armies made war upon us on the plateau yonder. But our means of warfare proved far more terrible and dreadful here than on our proper star. Thus have we unwittingly slain two of thy soldiers and frightened all the army. We have with us the means to kill them all, but we seek a peaceable life here for a brief time, that we may learn your ways and test your wisdom, when we shall be gone again."

"The Pharaoh could have better spared a thousand men than the bird which thy lightning hath killed. For are not his slaves as the plenteous grain of a rich harvest, while his birds are but as the fingers of his hands. If ye came but to learn, 'tis well ye know these wise men, though, since I came to Kem, their profession hath fallen somewhat into disrepute. I doubt not but they could learn far more from thee than thou from them, but they will not do it. Whatever they do not know is not true in Kem, but what they know continues true long after common men know better. Now, wilt thou explain to me the mysteries the soldiers have reported to us? But first tell us which of all the stars it is thou comest from."

"Know then, O Zaphnath, that we call our star the Earth, and in her wanderings she hath now approached so near to the great Orb of Day that her rays are paled by his brighter light; she sets with him, and shines no more by night. But yet a few days now, and she shall triumph even over him, and, entering on his glowing disc, she shall be seen at mid-day, obscuring his light and travelling as a spot across his glory."

The old men wagged their beards as the boy translated, but he sprang to his feet with no little excitement, and exclaimed,—

"Meanest thou that blue star with its attendant speck of white, which but a little while ago shone with great brightness as a Twilight Star?"

"That is the Earth, O Zaphnath, the world from

whence we came," I exclaimed; and the youth again threw off his wig and beard, and rushing toward me, pressed first his right cheek and then his left cheek against mine, and then against the doctor's.

"Then ye are most welcome to the land of Kem, and we shall be friends for ever. For ye should know that my mother was barren all the years of her life until this same Blue Star came to shine wondrously, even in the presence of the Day-Giver, before his setting. It was then, under the beneficent influence of this star, that she gave birth to me. And when the star paled and wandered again I tarried not in the land of my father, but came strangely hither, to be ruler in a great land which my people had never known."

When he had resumed his seat again, I said, "All that I have told thee shalt thou see come to pass, and through this Larger Eye, which we have made to pierce the deep of space, thou shalt see more clearly that the Blue Star is indeed a great orb, where many men may dwell, and after she hath passed the Day-Giver, she will appear as a bright morning star again to announce his coming."

"Why now, if this be true, then every one of these old men must die. For Pharaoh's laws provide that whatsoever wise man faileth to predict such an appearance, or predicteth one which doth not occur, must lose his life. These grey-beards, always jealous of me, have said that the Blue Star, which beareth my destiny, hath disappeared, never to be seen again. Now, when they are slain, Pharaoh shall appoint you to sit in their places. Ye shall reign jointly with Zaphnath if it pleaseth you, and ye may choose what seemeth good to you of everything that is in the land of Kem and in all the countries which pay tribute unto Pharaoh. And he will give you as wives all the women ye saw in Long Breath Park, and an equal part of all the slaves and women taken in war will he give you also. For hath he not bidden me treat generously with you, even to his tributary countries and half his women?"

"We come from a star, O Zaphnath, where men desire many things and are never satisfied. But of all the things thou offerest us, we wish not one. We make no peace unless these old men be left alive. We do not know this country or its people, wherefore we are most unfit to rule them. We wish no slaves, but will pay a hire to one or two good men, who may do our daily tasks. And as for women, we never choose but one, and then only when we know her well and find her equally willing."

"Then are ye come from a most strange star indeed! But I must tell thee that the laws of the Kemi forbid even to the Pharaoh, who hath the first claim upon all women, to take to wife a woman such as her whose hand thou clingest to so warmly. What findest thou in her whose dumb tongue could never tell thy praises, and if 'twere loosened, her mind would still be dumb and silent?"

"Who is this woman, then, whom thou sentest out to meet us? She alone hath had no fear, and hath greeted us in a friendly and a welcome manner. Had it not been for her, we might still have been loosening our thunder among your soldiers, or flashing this lightning in thy face!" I said, half drawing my long sword as I spoke.

"She is Thenocris, a poor, unfortunate maiden, dumb of tongue and mind," he answered. "In my country we would call her mute and senseless, but here among the Kemi they revere such ill-starred creatures, thinking that because they act strangely, and look not upon the world as others do, their souls must be turned within to the contemplation of hidden and spiritual things. They think such creatures know the secrets of the gods, and that the gods have made them mute, or speaking only silly things, lest those secrets be revealed. The people, therefore, give them alms, and suppose that they are effectual in intercessions with the gods. This girl went out at noon, as was her custom, to stand by the gate and ask alms. A soldier saw thee seize her hand and hold it strangely long, and he reported this to us. Whereupon these wise men with one accord decided that ye must have come for women, and we set about preparing a peace-offering of two thousand maidens for you in the Park. Afterwards there came another soldier later to say that ye had landed in the Park, pleased with our offering of the women. Then rose yon grey-beard and argued most wisely thus: That ye, being such strange creatures, had understood best what we understand the least; that thou hadst learned the hidden thought of this dumb woman by long holding of her hand; that, as ye had been friendly to her, she might be able to lead you unto us; and lastly, that it would be no breach of our laws if thou tookest this woman to thine own land and madest her thy wife; that if we could thus save our city, and the lives of the people, it would be wisdom to give her to thee, together with all the women in the Park. Then another grey-beard, wishing to share the credit for a wise idea, arose and insisted that it would be ill in us to keep the strange white animal, which one of the men found upon the plateau. We knew that ye must have brought this, for in all our land we have no four-footed thing smaller than the useful burden-carrying asses ye have seen. Wherefore, the wisdom of the grey-beards being now complete, we sent the dumb girl and the white animal out with the soldier, and they have brought you hither."

"So you have been falling in love with a queen of your own making, who is no more than a dumb idiot!" chuckled the doctor.

"Silence!" I shouted hotly, for I was unspeakably sorry for the poor girl. "There are softer, kinder words than those by which to call a poor blank soul that's born awry. The Kemi are quite right, for this girl, having no sense, has yet been wiser to-day than both of us and all these wise men." Then turning, I addressed the ruler in Hebrew :

"Thou shouldst know that in our land the seizing of the right hand is a salutation of friendship and welcome, much the same as the pressure of the cheek is here. We had vainly tried to signal to your soldiers that we were friendly, and when this woman stretched out her pretty hand I was pleased to seize it warmly. Call thou a soldier now and send her safely home. Let the white rabbit belong henceforth to her. She hath unwittingly been God's messenger in bringing us together. Mayhap she hath saved the lives of many of the people. Wherefore let them remember her, and henceforth treat her kindly. And as for those other women in the Park, bid them all return to their homes, and let it generally be known that there will be peace, and no further war. The terms of truce we will arrange with thee and with the Pharaoh somewhat later. We wish no gifts or offerings of peace. No more do we desire than that the Pharaoh shall entertain us for a season until we learn your ways, and then permit us to live quietly in this, your city, obedient to your laws, and pursuing such careers as our abilities may fit us for."

"All this that ye desire, and more, most gladly

shall be done, and a grand festival shall be appointed for this night to celebrate the peace. The Pharaoh will entertain you and his royal friends with feasting and with dancing, and the terms of the compact between us shall then be ratified."

At this point a grey-beard interrupted the young ruler, and a spirited conversation took place between them, after which the youth asked,—

"Tell me now, are there not many more such men as ye upon the Blue Star, who may come to wage a further war with us?"

"Have no fear for that," I answered. "The vessel in which we came is the sole means of bridging that vast space, and no more can come, unless indeed we bring them. But all of them shall keep the covenant we make with thee."

Then Zaphnath held a long consultation with the wise men, which ended by the summoning of three soldiers—one to take the woman home, another to carry the news of peace to the Park and to the people, and the third, as I supposed, to convey a message to the Pharaoh; but before the last was despatched, Zaphnath said to me,—

"Our messengers reported a third curious person with you, having a much larger body and long moving horns. What have ye done with him? Is he left in charge of your travelling house?"

Then I explained this circumstance to them, as well as the incident of my smoking, which I promised to repeat at the banquet in the evening.

After hearing this they dispatched the third messenger.

"We have heard, not only that ye breathed smoke and carried flames in your limbs, but that your flesh was of iron, invulnerable to arrows; that ye were stronger than birds, and carried the thunder and lightnings of the gods with which to kill; and that ye were able to walk through the air as well as on the ground."

"'Tis true we are stronger than any birds upon our proper star, and that we kill with a thunder and a lightning. Our flesh is tougher and more solid than thine, yet 'tis not of iron. But tell me, what knowest thou of iron?"

"'Tis a rare, precious metal which we coin for money, but I see thou carriest much of it. Thy thunderers are made of it."

"And hast thou no metal, bright and yellow, such as this?" I asked, exhibiting my gold watch.

"In truth, the Pharaoh alone is able to possess such riches, and in all the land of Kem there is no such huge lump of it as that!" he exclaimed in wonder, while the sleepy wise men opened their big eyes.

"We have within our belts many coins of this, which we may barter with the Pharaoh for things more plenteous here."

"Are ye travelling traders then, or what were your occupations on the Blue Star? Were ye warriors, rulers, wise men, or owners of the soil?"

"My good friend here hath been a wise man, as thou must know from his grey beard," I answered, smiling at the doctor. "He hath been a teacher of knowledge to the people, and it was his superior wisdom which contrived the house in which we travelled hither."

"But hath it not been a folly to teach wisdom to the people? When they have learned, the wise man turneth fool! Wisdom groweth ripe by being bottled, but whoso poureth it out for every thirsty drinker wasteth good wine upon gross beasts!"

"In its youth our star held to these opinions, but now it teacheth wisdom to every child, and in this manner we have made progress into many things not even dreamed of here. As for my own profession, I have been a dealer in wheat, the breadgrain of our star. Hast thou here such a small grain growing at the bearded end of a tall straw?"

"In truth, the land of Kem raiseth so large a store of such a grain as to feed all the surrounding countries! Our greatest traffic is in this wheat. Hast thou not seen the green fields of it lining the banks of the Nasr-Nil, until the sight tires following it? This season there cometh such a crop as Kem hath never seen before, and for six years we have been blest with its plenty——"

Here he was interrupted by the hurried return of the third messenger, who addressed him in excited tones. As the Kemi use no gestures, and but little facial expression in their conversation, I could not guess the import of his message. Therefore when it was translated by the youth it was all the more surprising.

"The soldier saith that a certain curious man of Kem, anxious to explore thy travelling house, ventured within it, when presently it rose and sailed away with him far out of the city, and was lost from sight in the red distance!"

This was an unforeseen, stupefying development. I left the doctor to guard our things, and rushing out I leaped the courtyard wall and ran with all haste to the Park. The projectile was gone! No sign or trace of it was anywhere to be seen. Willingly or not, we were henceforth chained to Mars!

CHAPTER VI

The Iron Men from the Blue Star

RETURNING from Long Breath, I could not but notice the entire subsidence of the terror, which had previously been so marked, and the general signs of rejoicing which were now taking its place. It was easy to see that I was an object of absorbing interest and busy comment. No one pointed the finger at me, for that rude gesture was as unknown as it was unnecessary. The mere turning of a great pair of eyes quickly in my direction was an indication, significant enough, that I was being denoted.

I now understood the more composed behaviour of the women. They were accustomed to the idea of being taken in war, and never suffered slaughter or hardship thereby, but merely a change of masters. As they now left the Park they eyed me curiously, as if wondering from what sort of new master they had escaped. I imagined I could detect some signs of disappointment among them, at being cheated out of a trip to a new star or being dismissed from the service of a god. Occasionally one of them would incline her head gently to the right to meet her rising hand, in a dignified salutation. I approached one of the fairest of these and extended my hand. She seemed rather surprised, but calmly placed an iron coin in my palm! Evidently I must make haste to learn the Kemish salutation, or I would pass for a common beggar! My hand certainly did look hard and brown, compared with her perfectly white and transparent skin, through which the blood suffused the beautiful pink flush of life. But even if a hotter sun had scorched and tanned my hand, it did not look as dark and tough as the coin, although the soldiers had spread the report that our flesh was of iron.

The chief business activity in the city seemed to be the transporting from the surrounding country of an endless number of fibrous bags filled with the bread-grain. I saw some of these bags open in the shops, and the grain was shaped like wheat, but as large and less solid than a coffee berry. Trains of asses bearing these bags were seen in every street and entering by every gate. Each train of fifteen or twenty asses was driven by a sandalled Martian, wearing the spread bird-wing which seemed to denote the Pharaoh's service. The animals had the lazy, sluggish, plodding habits which I expected, and in these respects their driver differed very little from them. He gave an occasional long hiss, followed by a jerky grunt, which sounded like "sh-h-h-h, kuhnk!" and was evidently intended to hurry the animals, but it served them quite as well as a lullaby. These drivers, who doubtless had just been hearing stories of me, were a little surprised at coming upon me so soon, but looked me over deliberately, as if calculating how much iron money I would make, if there were no waste in the coinage!

But I hastened back to the doctor at the Palace, being obliged to leap the courtyard wall again, for I was not acquainted with the signal to command the Terror-birds. He expected no other report of the projectile than the one I brought.

"The only hope is that the meddling Martian may have turned in but one battery," he said. "In time this will exhaust itself, and the projectile will tumble back upon Mars. If it should strike in the water, it may not be shattered, but of course it might be submerged. The chances that we will ever see it again are extremely remote. If it should be discovered anywhere on the planet, it would probably be coined up into money, and the fortune of the Pharaoh would hardly buy us iron enough to make another. Well, the unexpected always happens. It was a fatal mistake ever to have left it."

"If it is gone for good," I answered, "let us hope that this planet may suit us better than the Earth, anyhow. We are certain of an easy existence here at least. One shield will coin into money enough to supply our wants a long time. If we had not been so dreadfully secretive on Earth, perhaps some one, infringing our ideas, might have built another projectile and sent a relief expedition!"

Preparations for the banquet were rapidly being made about the Palace by men servants. We saw no female servants, and we learned afterward that they did no menial work, except the serving of the meals, which was rather an artistic duty.

We were conducted to two large ante-chambers, adjoining the banquet room, where we deposited our armament and proceeded to make ourselves at home as well as we could. The rooms were gloomy and poorly lighted, but a great number of servants were busy waiting upon us, and one presently brought in four portable gas-burners, placing them in a circle about my head as I reclined on a large pillow of soft down, laid on the floor. These burners thus furnished both heat and light, and nearly all the rooms were thus lighted and heated throughout the day. They had windows and a very thick, coarse, translucent but not transparent glass in them. But as the sunlight was never strong, rooms were rarely ever light enough for comfort without the flames of gas.

This was my first acquaintance with Martian gases, which I soon found to be very numerous and various in use. On the other hand, very few liquids existed. The atmospheric pressure was so low that what might have existed normally as liquids on Earth, took the form of heavy gases here. In every case they were heavier than the air, so that they remained in vessels just as a liquid would have done. The four lamps were made of reeds and shaped like the letter U. The right-hand side of the U was a large vertical reed, connecting neatly at the bottom with a very much smaller reed forming the other prong and terminating at the top in a tip of baked earth, turned downward, so that it would discharge the gas away from the lamp. A light stone weight was fitted to slide neatly down the large vertical tube in which the gas was stored, and thus force the gas up to the burner in the smaller tube. If a brighter light was desired, a heavier weight was put on, and to extinguish the light it was only necessary to lift the weight, which cut off the supply from the burner.

While lying on the downy floor-cushion, I was strangely annoyed by the faint and distant howling of a dog. It seemed to come from the banquet room adjoining mine, or from the doctor's room on the other side. I called in the doctor, who said he heard nothing and had seen no dogs on Mars. He tried to make me believe it was a fancy of mine. But presently when a servant entered, he seemed to hear it instantly, for he turned quickly about and left, but it was fully half an hour later before the plaintive howling ceased.

THE IRON MEN FROM THE BLUE STAR 225

"These Kemish people have better ears than we have," I remarked to the doctor.

"Yes, both their ears and eyes are much better suited to the conditions of fainter light, and higher, thinner sounds. There may be music at the banquet to-night which we cannot hear at all in some of its notes."

"If there are no foods whose delicate flavours we fail to taste, I shall be able to get along quite well. I am extremely hungry, and quite ready for a change of fare." We had only eaten a hasty lunch when we had re-entered the projectile at Long Breath to await the return of the soldier.

Zaphnath himself came to conduct us to the banquet room, and we were much surprised at its dark and gloomy character. The entire vast enclosure had but twenty-one flickering firebrands, suspended overhead and in front of us, to furnish light. There were no tables or chairs, no flowers or decorations, no sign of anything to eat. Other guests were moving about through the semi-darkness to their places, seemingly without inconvenience. I was whispering to the doctor that I would need eyes of much greater candle power to enjoy the function, when we arrived at our places. A double row of comfortable cushions ran along the edge of our floor, where it seemed to sink to a lower terrace, whence we could hear the indistinct hum of women's voices. Zaphnath took his seat on a raised cushion in the middle of the row, and

P

motioned me to the cushion on his right and the doctor to his left. Eighteen other guests now reclined upon their cushions to left and right, so that we were all arranged in a direct line, facing the lower terrace whence came the feminine buzz. Directly opposite each of us was an empty cushion, but no table.

I was wondering at it all when the fire-brand farthest from me suddenly exploded a great flaming ball of fire, and we all sprang to our feet. From the terrace below came a grand burst of reed music, a swelling chorus of women's voices, and then each firebrand in quick succession exploded a burst of flame, which floated down toward the dancing women, but expired above their heads. I soon saw that these white fire-balls, which continued in quick succession throughout the banquet, and afforded us a glorious if a somewhat appalling light, were caused by the successive discharges of small volumes of heavy gas from twenty-one reed-tanks in the comb of the roof, one above each of the fire-brands. When the discharged gas had floated down to the fire-brand beneath it, there was a quick, bright explosion, and the flame sank menacingly toward the women below.

The burst of music, the chorus of huzzahs, and the flashing forth of light, proved to be a welcome to the Pharaoh, who was standing proudly on his great throne opposite us, across the terrace and somewhat higher, whence he could look down upon the dancers and singers. He wore a crown of thin iron, surmounted by a golden asp. His elaborately curled wig did not conceal his ears, from which large golden pendants hung almost to his shoulders. His own beard was waxed and curled, and trimmed to the shape of a beaver's tail. His dress is best described by calling it a feather velvet, edged with flaring wing and tail plumes of iridescent colours. In this feather cloth there was none of the rough, gaudy show of the savage, but a discriminating, tasteful blending of colours and harmony of design, imitated from the beauty of the bird itself.

Grouped about him on the approaches to his throne were one-and-twenty of his favourite women, beautifully dressed in feather textures, with the curved neck and head of a bird surmounting their brows. But their costume was scant and simple compared with that of the dancing girls below us. They wore a wonderful head-dress, composed of the entire body of a small peacock. The head and neck were arched over the forehead, the back fitted tightly, like a hat over their head, the drooping wings covered their ears, while the fully spread tail arched above their head in its wonderful opalescence. Much of the snowy whiteness of their backs and breasts was bare, and a downy feather ribbon circled the necks, wrists, and ankles. A two-headed iron serpent with golden eyes clasped the upper arm and gartered the knee, but no jewels of any kind were to be seen. All the dancers carried long decorated reeds, which they flourished wondrously, and with which occasionally they executed the most surprising leaps. While there was a stateliness about their movements, there were also the most startling acrobatic surprises, made possible by the feeble gravity.

The singing women, or what might be called the chorus, were in twelve sets, each group clad in a different colour or design of feather-silk. Their head-dress, while composed of the entire body of a bird of plumage, lacked the flamboyant tail of the peacock. The music was weird and whimsical, as there were neither stringed nor brass instruments. It was made wholly by women playing upon a vast variety of drums and reeds. There were all sizes of whistling reeds or flutes; several of these of different lengths were grouped into one instrument like the pipes of Pan; a series of long hollow reeds, when rapidly struck, gave forth a marvellous cadence; while groups of small drums, of different size and tensity, gave curious tones. The whole effect was weirdly eloquent, rather than racy or exciting.

When the burst of welcome was ended, Zaphnath stretched forth his hand and exclaimed, first to us in Hebrew, and then in Kemish,—

"O Pharaoh, whose power and wisdom from all the Pharaohs have descended, behold, I bring unto thee these two iron men from the Blue Star, who, though excelling in the arts of war, are yet pleased to come out of the ruddy heavens to practise peace amongst us!"

And thus did Zaphnath translate the Pharaoh's response to us:—

"Unto Ptah, the Centre of Things, to whom the myriad stars of the heavens are but ministering slaves, I, Pharaoh of Kem, do give you welcome. Whatever pleaseth you in the largeness of this rich land, or in the matchless beauty of our women, shall be unto you as if ye had owned it always."

Whereupon the other guests turned toward us with the right hand upon the cheek, and we awkwardly attempted the same salutation. Then a group of the singing women, twenty-one in number, tripping to the weird music, came up the steps which led to our floor, carrying covered dishes. At the top they turned and saluted the Pharaoh, and then took their places, one upon each of the cushions opposite us. Before uncovering the dishes they took me a little by surprise, by bending forward and pressing their warm, pink cheek against the right cheek of the guest they were about to serve. My maiden unconsciously shivered a little, for my cheek must have felt cold, even though my surprised blushes did their best to warm it. Her dish, when opened, contained nothing but flowers, waxy white, but emitting a delicately sweet perfume. She held them toward my face, and presently breathed gently across them, as if to waft their perfume to me. Then scattering them about my cushion, she pressed her left cheek to mine, arose and tripped down the steps again. There was a modest self-possession about her which enchanted me, and I hoped she would soon return bringing something more substantial.

But another group of maidens, differently clothed, had already begun to mount towards us with earthen goblets and reed-pitchers, which looked as if they might contain wine. Dropping on her knees on the cushion before me, this maiden saluted me as the other had done. Then sitting gracefully before me, she tipped her reed pitcher toward the goblet, and poured out apparently nothing! But, watching the others, I saw them carry the goblet to their lips and draw a deep breath from it, while tipping it as one might a glass of wine. I did the same, and inhaled a deep draught of stimulating, wine-flavoured gas, which, when I exhaled it through the nostrils, proved to be deliciously perfumed.

"I have heard of some poets who could dine upon the fragrance of flowers and sup the sweetness of a woman's kiss, but I am hungry for grosser things," I whispered to the doctor.

"There are ten other groups of these serving maidens to come up to us," he replied. "They will certainly bring us something more tangible before it is over. Meantime, while we are in Kem, let us imitate the Kemish;" and I must say he was succeeding remarkably well.

The next maiden who tripped up toward me was wonderfully beautiful and most becomingly dressed. I was a little disappointed that, upon taking her place on the cushion in front of me, she omitted the salutation the others had given. However, she carried a small flask in her right hand, which she placed near my mouth. Then opening the top of it slightly, it jetted forth a deliciously perfumed fine spray, which moistened my lips. Waiting just a moment for me to enjoy the perfume, she then pressed her pretty cheeks in turn against my lips, until they were soft and dry. This was the nearest approach to a kiss which I saw among these people, and I learned it was given always just before eating solid food. The plate she carried to me contained small morsels of fish, served upon neat little wheaten cakes. There was no knife, fork, chopstick, or anything of that kind, but each little cake was lifted with its morsel of fish, and they were together just a delicate mouthful. This maiden quite took my fancy, and I watched her evolutions and listened for her voice in the chorus during the rest of the banquet, for she had no more serving to do.

After this course Zaphnath arose, and waving to the music and singing to cease, he thus addressed the Pharaoh :—

"It doth appear, O Pharaoh, that these visitors of ours come from a strange, small world, where, though much is done, but little is enjoyed. At thy bidding I have offered unto them all the luxuries of Kem, such as our people strive all their lives for, and dying still desire; but they wish no gifts or presents. Like slaves they only wish to work, but at some noble, fitting occupation. This younger man, whose wondrous learning hath taught him to speak even the tongues of other worlds, hath been a great handler of grain upon his proper star, and for him the fitting occupation is not far to seek. Thou knowest how the gathering of thy bounteous harvests hath distracted my own attention from weightier matters; wherefore, O Pharaoh, I do entreat thee to put into his charge the labour of gathering, storing, and distributing all thy harvests; and as a fitting compensation, let him have one measure of grain for every twenty that he shall gather for thee."

Nothing could have suited my wishes and abilities better, and my pay on Earth had been only one measure in five hundred. The Pharaoh's reply was thus translated to us,—

"The gods put into thy mouth, O Zaphnath, only the ripeness of their wisdom, and Pharaoh granteth thy requests ere they are uttered. But what desireth the wise man?"

To this I made answer for the doctor,—

"When thou knowest his wondrous wisdom touching many things, O Pharaoh, thou mayest think fit to give him a place among thy wise men, where they may learn from him and he from them.

Will it please thee to send a slave for the Larger Eye and have it placed by yonder window, and he will presently show unto thee many of the wonders of the starry heavens that are hidden beyond the reach of man's unaided vision."

While two slaves were despatched in charge of a soldier to bring the telescope, we were served with a highly-sparkling, gas-charged wine, which further whetted my appetite. Then came another maiden with a small roast bird, neatly and delicately carved, and each tempting piece was laid upon a small lozenge of bread. I never ate anything with more relish.

There was an excited buzz among the women, and the Pharaoh himself was visibly affected at the sight of the telescope, whose burnished brass was evidently mistaken for gold. The doctor mounted it upon the backs of slaves near a high window, whence there was a good view of the heavens, and signalled to me to explain its uses.

"O Zaphnath, wilt thou make known unto the Pharaoh, and these, his guests, that the wondrous value of this instrument lieth not in its bright and glistening appearance, but in the farther reach and truer vision of the heavenly bodies which it affordeth us. With this we ascertain all and far more than yon monstrous Gnomons tell thee; we learn the periods of the day, the seasons of the year, and vastly more than our common tongue hath words to tell thee of. Tell me, what callest thou yon risen orb, which hasteneth a rapid backward journey through the heavens?" I asked, indicating the full disc of Phobos.

"That is the Perverse Daughter, sole disobedient Child of Night, whose stubborn, contrary ways are justly punished by her mother. For she must draw a veil across her brilliant face for a brief period during every hasty trip she makes."

"Behold her, then, just entering upon her punishment!" I exclaimed, for the regular eclipse was just beginning. "Look! and tell us all thou seest."

"I see a glorious orb, far larger than the Day-Giver and very near to Ptah! But it is the Perverse Daughter, grown larger and come nearer, for she alone knoweth how to draw the veil of night across her face like that. Now she hath fully hidden! It is most wonderful, O Pharaoh!"

"Be not deceived by mere appearance, O Zaphnath," replied the Pharaoh. "All that thou seest may be contained within the thing thou gazest into. 'Tis true, the Perverse Daughter hath drawn her veil, but be thou sure thou seest what is beyond and not merely what is within."

As soon as this was translated to us, the doctor focussed the telescope upon the Gnomons, which were just visible over the edge of the plateau, and I said,—

"Look now again, and behold all the familiar features of the landscape, the plateau yonder and the ponderous Gnomons, which could never be contained within this little enclosure."

"'Tis all most true, O Pharaoh, and with this little instrument thy reign may be more glorious, and come to greater wisdom, than any of that long line of Pharaohs, whose toiling slaves have built the towering Gnomons. Let this grey-beard be made chief of all thy wise men; let the others teach him our language and make him acquainted with all our monuments and records; also command them to record most faithfully all the wonders which he is able to reveal. Mayhap he may be able to write thy name among the stars of night, to shine for ever, instead of upon the crumbling stone which telleth of thy ancestors!"

"O men of Kem," replied the Pharaoh, addressing the other guests, "hear ye the wisdom of Zaphnath, which cometh with the swift wings of birds, while thy halting counsel stumbleth slowly upon the lazy legs of asses! What Zaphnath asketh hath already been decreed touching these two men from the Blue Star, provided only that they live peaceably among us obedient to our laws."

We assured him of our obedience and our best efforts to discharge our new duties, whereupon the feast continued. Courses of small birds' eggs and of fruits and confections were each served by a separate group of maidens. When the feast was finally completed, I turned to Zaphnath with my cigars and said,—

"In our travelling house I brought with me many such things as these and others of a smaller, milder form, which might delight the women; but now that the house is gone, I have but three, one of which wilt thou send to the Pharaoh, one keep for thyself, and the other I will smoke to show you the manner of it. There is naught to fear about them; your taste for heavy vapours will have prepared you to enjoy the warmth and fragrance of this peculiar weed."

A servant came to carry the one to the Pharaoh, and I struck a match upon the stone floor and held the cigar designed for Zaphnath in the flame. Then I touched the flame to my own, and puffing gently, I asked Zaphnath to do the same. When I saw that his custom of inhaling gases led him to breathe in the smoke, I puffed very slowly and gently, until he should become accustomed to it. When Pharaoh saw that it did no harm to Zaphnath, he lighted his own and inhaled the smoke in long draughts with evident gusto.

"How sayest thou, O Zaphnath," he said at last. "Is not this warm vapour most stimulating? It is a treat worth all the rest of the banquet. Continual feasting hath made the luxuries of Kem to pall upon me, but this hath novelty and comfort in it. If, indeed, there were many of these in thy travelling house, my slaves shall search all the width and breadth of Ptah, until it be found."

The music now burst forth again in new volume,

THE IRON MEN FROM THE BLUE STAR 237

and the singing girls went through a new evolution, which broke up their groups and formed twelve new ones, containing one girl from each of the previous sets. Then the entire number began ascending the steps together, and I noted that those approaching me were the twelve maidens who had served me during the banquet. They came and circled around me, and presently stopped with their hands upon their cheeks in salute. The other groups did the same to the guests they had served, and each guest selected a maiden by saluting her upon the cheek, whereupon she left her circle and took her position upon the cushion opposite him. Zaphnath, seeing that we did not understand this ceremony, explained it to me.

"It is an ancient custom with the Pharaoh to present each of his guests with a living reminder of the occasion and his hospitality. Wherefore he desireth thee to choose which of the twelve serving maidens hath pleased thee best, and he will give her to thee, to be always thy maidservant."

I translated this to the doctor, and watched him curiously, with an inquiring twinkle in my eye.

"Let us accept them, and bestow their liberty upon them," he said.

I immediately chose the third maiden, who had pressed her pink cheeks to my lips, and when she came to sit opposite to me upon the cushion, I spoke to her through Zaphnath,—

"Thy ways have pleased me, but upon my star

we do not think it proper to own any slaves. When we know well-favoured and graceful women, such as thou art, we prefer to be their slaves, rather than they ours. If I could take thee with me to the Earth, the laws there would set thee free to do whatever pleased thee best. Wishest thou that I make thee free here?"

She was evidently surprised when Zaphnath put this question to her. She replied in a sincere and pleading tone, but her words astonished me,—

"Whatever the dark Man of Ice wisheth, I will do. I know not why he hath asked what I desire. He speaketh of freedom, but I beseech him not to send me back to that! I was born an unhappy and masterless maiden, and many years I struggled and laboured for a miserable existence. I drove asses, gleaned in the fields, and did the menial work of men. But I felt I was fit for better, nobler things. At last, I heard that the armies of the Pharaoh were coming to my land, and I took heed of my appearance, put on my neatest feather clothing, and went to throw myself before the soldiers. They were pleased with me, and brought me to this city, where fortune favoured me, and Pharaoh, looking over all the women whom the soldiers brought from the wars, chose me, with many others, to join his household. And here in the Palace for a few years I have been happy and well cared for. I pray thee do not turn me out again; do not degrade me to the labour and misery of freedom. Even the beasts have masters! They are housed, and fed, and cared for; why should I then be cast out and left to drudge or beg?"

"Doth she mean this?" I exclaimed. "What then is the chief aim of women in Kem? What is the highest state to which they may aspire?"

"'Tis a strange, simple question!" he answered. "There is no greater blessing for a woman than to belong to the household of the Pharaoh. Here they are delighted with constant music and dancing; their beauty is cultivated and heightened by rich and tasteful clothing; and their charms and graces may win for them a selection as one of the one-and-twenty favourites of the Pharaoh. What they fear most is being chosen and carried away by guests whose palaces and ways of life are less luxurious than the Pharaoh's."

"Why then, as we have no palaces and wish no slaves, it were best to return these maidens to the Pharaoh if they will be happier and better cared for here than anywhere else in all the land of Kem," I said to Zaphnath.

"This age is not ripe for the grand idea of freedom which dominates our own," remarked the doctor, as we returned the grateful maidens to the constant delights of an ornate and sensuous slavery.

CHAPTER VII

Parallel Planetary Life

I WAS sleeping soundly on my deliciously soft heap of downy pillows, when in the early morning I was awakened by a pounding on the door of the ante-chamber. As one always wakens from a sound sleep with his most familiar language upon his tongue, I cried out in English, "Who's there?" The doctor answered, wishing to be let in. I fumbled about in the darkness sleepily, and opened the door, and he lighted two of my gas-lamps with the one he carried. He looked rather tired and worn.

"I am possessed by a tyrant idea, which will not let me sleep," he said. "I must get rid of it before morning. Come, get your senses about you, and listen to me," he commanded, as I yawned and rubbed my fists into my eyes, blinded by the sudden strong light.

"If you think I can sleep with it any better than you can, out with it," I answered.

"How does it happen that a young Hebrew is ruler over all these people?" he demanded.

"Do you lie awake thinking up conundrums?" I ejaculated.

"On Earth, what notable Jews have been rulers over a great people not of their own race?" he continued.

"Disraeli in England, Joseph in Egypt, and—well, that is all I can think of just now."

"Perhaps that is enough. Egypt was the greatest grain-raising country in Joseph's time, wasn't it?"

"Yes, of course," I answered. "And Joseph's rule began with seven years of most wonderful crops."

"Zaphnath told us this morning that the seventh great crop, and the most plenteous of all, is now growing," he interrupted.

"What has that to do with Joseph? We are not on Earth, but on Mars. Have you been dreaming? Zaphnath is—— But, by the way, Joseph's Egyptian name was Zaphnath-paaneah, meaning a revealer of secrets! When I heard that name this morning, I thought it was strangely familiar. Pharaoh called him that when he appointed him ruler, because he had interpreted his dream," I said, just realizing the very peculiar coincidence.

"You are as good as a Bible!" cried the doctor. "Perhaps you can also remember by which of Jacob's wives Joseph was born?"

"Of course I can. He was the first son of

Rachel, the wife whom Jacob really loved, and worked fourteen years to secure."

"But how could he have ten older brothers, if he was Rachel's first son?" he demanded, a little perplexed.

"They were all the sons of her sister Leah and her handmaidens. Rachel was barren all her life until Joseph was born," I explained.

"And Zaphnath said this morning that his mother was barren all the years of her life that the Blue Star wandered. He also called himself revealer of God's hidden things."

"Yes; and it struck me as peculiar at the time that he said of '*God's,*' not of '*the gods*,'" I reflected. "Evidently he thinks there is but one God. The whole matter is altogether peculiar."

"Here are the facts," replied the doctor. "Listen to them attentively. We have dropped down into a civilization here upon Mars which coincides in every important particular with that of the Ancient Egyptians on Earth. They are great builders, erecters of monuments, raisers of grain, polygamists, and they now have a young Hebrew ruler, corresponding in every important respect with Joseph. We chance to have arrived during the seventh year of plenty of Joseph's rule. Grain abounds; the soil brings it forth 'by handfuls.' It is, 'as the sand of the sea, very much,' and the Pharaoh, probably at the suggestion of his young ruler, is storing it up——"

"By all the Patriarchs!" I interrupted. "They are running a wheat corner, and I didn't know it! Go on, go on!"

"These are all very singular coincidences with a history which was enacted many thousands of years ago on Earth. Now, how can you explain their strange recurrence here?" he queried.

"How should I know? I haven't been lying awake! How do you explain them?" I asked, full of interest.

"I have tossed on my pillows in there for three hours evolving a theory for it. If it is correct, our opportunities here in Kem are simply enormous. Now listen, and don't interrupt me. The Creator has given all the habitable planets the same great problem of life to work out. Every one of His worlds in its time passes through the same general history. This runs parallel on all of them, but at a different speed on each. The swift ones, nearest to the sun, have hurried through it, and may be close upon the end. But this is a slow planet, whose year is almost twice as long as the Earth's, and more than three times that of Venus. The seasons pass sluggishly here, and history ripens slowly. This world has only reached that early chapter in the story equivalent to Ancient Egypt on Earth. We have forged far ahead of that, and on Venus they have worked out far more of the story than we know anything about. If Mercury is habitable yet, his people may have reached almost the end, but it is most probable that life has not started there; when it does begin, it will be worked out four times as rapidly as it has on Earth."

"Then a seven years' famine will begin here next year, and I am in charge of the world's entire wheat supply!" I gasped, almost overwhelmed by the speculative possibilities which this unfolded.

"It is not likely that there will be more than a general similarity of the history. But Zaphnath has told us that this is the seventh year of plenty. If the famine begins soon, it will be fair to suppose it will for about seven crops. In its later developments the entire history may change when the crucial period comes, and have a very different outcome. But we are now almost at the beginnings of civilized history. Joseph, the first Jew in Egypt, is a ruler here, and your entire race must follow him hither, and pass through a miserable captivity. Even if you remained here all your life, you would not last that long; but upon the later doings of your people and their treatment of the Martian Messiah, when He comes, depend the future conditions of this planet. Will it be different then from the Earthly story? It is an extremely interesting theory to follow to the end, but that would take thousands of years, and we are concerned with the present."

"Doctor, if this theory be true, then we are nothing short of prophets here!" I exclaimed, still struggling with the wonderful bearings of the idea on our personal welfare.

"In a general way we are prophets, but Zaphnath has forestalled us on immediate matters. Let us keep our own counsel as to any foreknowledge. If we disclose it, we may suddenly lose our opportunities, and, besides, we shall be powerless to change history here in any important respect."

"I might prevent Zaphnath from bringing all Israel down into Egypt, and thus save them from that captivity," I exclaimed.

"Then you would forestall a Moses, and prevent the miraculous deliverance of your people, and all the paternal care which God bestowed upon them during that time. You will never be able to do this. Zaphnath is in the way. He is headstrong and wilful. He is an active thinker and a hard worker among a race of idlers, who live only to enjoy the fulness of a rich land. He knows the greater activity and industry of his own people, and he will wish to make them masters of this goodly land. I will warrant that his head is full of plans at this very moment for bringing his old father and all his race down here to give them important places. See how readily he gave the keystone of the whole situation to you. It will pay you better to keep on good terms with him. Instead of trying to change the situation, let us make the best of it as we find it."

"Well, I must say the present situation is

attractive enough to me," I said, and then inquired, "How many gold coins have you, Doctor?"

"I have only a hundred half eagles and a little silver coin," he replied; "and I wish to be very sure of the correctness of my theory before I undertake any speculations with that."

"Nonsense! What is money for, but to double, and then to double the result again!" I exclaimed. "You work out this great theory, and then fail to grasp its commercial importance to us. You and I will embark in the grain business, with our entire stock of gold, the first thing in the morning. We have iron enough to live on."

"I didn't come here to go into business," he answered. "I have a grand scientific career to pursue, and last night's appointment puts me in just the position to carry it out."

"Go ahead with it then, but invest your gold coins in my enterprise. I will manage it all," I said, reaching for my belt under my pillow. "I have here three hundred eagles and one hundred double eagles,—five thousand dollars in all. I scarcely need your five hundred dollars, but I don't wish to see you left out, and buying bread of me at a dollar a loaf in a short time. Gold must have an enormous value here, considering the small amount of it used as ornaments in the Pharaoh's household, and the general currency of iron money. Three of these double eagles would make a pair of ear pendants equal to his. I wonder how he would like to have pure gold bracelets on all his women instead of those rough iron things? And wheat must be cheaper than dirt after seven enormous crops. I will buy all the grain he has to sell before tomorrow night! Even if your theory is all wrong, we can't lose much."

"That is all very well, but we may as well be sure," he replied cautiously. "You can find out much by a few discreet questions to Zaphnath in the morning."

"The trouble about the whole matter is, that I will be obliged to do business through him altogether until we learn this language. Come, you must contribute your share. I have furnished the Hebrew, you must learn the Kemish at once through these wise men. But I can't wait for that. I will make Zaphnath teach me the necessary shop words and stock phrases for carrying on the grain business to-morrow. I can't perform my new duties unless he does that."

However, the doctor did not respond wholly to my new enthusiasm. He was sleepy, and retired yawning to his own room to get the rest which had evaded him. But I lay and tossed on the pillows, revolving a hundred plans, and feeling anything but sleepy. Presently I thought of a scheme, which would demonstrate whether there was anything in the doctor's theory. I knew it would just suit him, and I sprang up and knocked gently on his door, saying,—

"I have it, Doctor. Here is the very idea!" There was no answer, so I knocked louder and listened. I heard him breathing heavily in deep slumber. After all, the morrow would do for ideas; just then he needed sleep.

CHAPTER VIII

A Plagiarist of Dreams

BEING unable to sleep, I arose early to get the refreshment of a morning walk. I passed quietly through the next room, where the doctor was still sleeping soundly, out into the courtyard. I was scarcely outside when I heard a familiar, excited barking, and Two-spot ran across the open space toward me as fast as his four short legs and his very active tail would carry him. His frantic jumping up toward me was extremely comical, for he sprang with more than twice the swiftness I was accustomed to seeing, almost to a level with my face, but he fell very slowly to the ground with only one third the speed that he would have fallen on Earth. He could jump, with almost the agility of a flea, and yet he fell back deliberately like a gas ball. He was evidently enjoying his muscles as much as I had mine. When he made a particularly high jump, I caught him in my hands and patted him fondly.

"So you didn't fly away with the projectile? Or, did you go with it, and is it safely back again, somewhere? How I wish you could speak my language and tell me all you know! These different tongues are a great bother, aren't they, Two-spot?"

He answered me volubly, but apart from the fact that he quite agreed with me, I could not understand his message. Had I been able to, it might have made a very great difference to me.

There was a beautiful, filmy snow on the ground, which had fallen during the night. It was scarcely more than a heavy hoar frost, and as the sun sprang up without any warning twilight, the snow melted and left the surface damp and fresh. As I afterwards learned, this thin snow fell almost every night of the year, except for the warmest month of summer when the grain ripened. There were hardly ever any violent storms or quick showers. The thin air made heavy clouds or severe atmospheric movements impossible. But the coolness of night, after a day of feeble but direct and tropical sunshine, precipitated the moisture in the form of those delightful feathers of darkness. I also learned that the months were distinguished by the time of night when this snow fell; for it was precipitated directly after sunset in the winter, but gradually later into the night as summer advanced, and finally just before daybreak. The month in which none fell at all was midsummer, of course. It had scarcely finished falling this morning when I came out into it.

I sprang to the top of the wall, and was watching

the quick rising of the Sun, and enjoying the sensation of looking fixedly at his orb without being dazzled, when I noticed that there was a dark notch in the lower left-hand part of his disc! Soon after I distinguished, somewhat farther in, a faint and smaller dark spot. This must be the beginning of the double transit of the Earth and the Moon! I experienced a sensation of joy in finding the home planet again. I confess it had given me a curious shock not to be able to see it in the heavens. It was more comfortable to have it back in the sky again, and at last I knew just where we were in the calendar. On Earth it was the third day of August, 1892. The summer there was at its height, and all my friends were as busy and as deeply immersed in their own affairs as if their little spot had no idea of coquetting with the Sun. Possibly a dozen pairs of studious eyes out of the teeming hundreds of millions on Earth were turned Marsward. This led me to wonder what all-absorbing topics of sport, politics, or war may fill the minds of the possible million people on Venus, when the Earth is so much excited over one of the infrequent and picturesque transits of that planet across the Sun.

But the doctor and Zaphnath must know of this! I hastened into the ante-chamber and called out,—

"Come, get up! I have already discovered two very significant things this morning."

"What are they?" he asked wearily between yawns.

"Two-spot and the Earth!" I exclaimed. "The former crossed my path in the courtyard, and the latter is just now crossing the Sun. Where is the telescope? quick!"

The doctor was not long in propping it up by the east window, and I went to look for a servant. By repeating the word "Zaphnath" several times, I made him understand that we wished the attendance of the young ruler, and he started for him.

By this time the notch was almost a complete circle of dark shadow within the lower edge of the Sun. The smaller spot, one-fourth the diameter, was forging ahead like a herald to clear the way. Zaphnath soon arrived, for he lived in another part of the Palace. He quietly pressed his cheek to mine, but in my excitement I had seized his hand, and with a pressure which must have hurt his shrinking flesh, I exclaimed,—

"This is the day of thy greatness, O Zaphnath, for, behold, the Blue Star is already upon the face of the Day-Giver!" I led him hastily to the telescope, and explained to him that the smaller forward spot was caused by a moon like Phobos, and that the Earth was really a round ball, like the Sun. He looked intently for a long time, and then turning about to me he said,—

"It is well ye left just when ye did, for the fire of the Day-Giver hath by this time burned every living thing upon your star! See how she hastens through his hot flames."

I attempted to explain that the Earth was more than twice as far from the Sun as she was from us; but he believed the evidence of his eyes, and I had to give it up in despair.

"I pray thee, bring this Larger Eye to the Council Chamber. I must summon all the wise men at once to behold this wonder. How long will it continue?"

The doctor told me it might last almost two hours; but I found it impossible to convey any idea of this period of time to Zaphnath, until I told him that it would continue half the time of the crossing of Phobos, who had just risen dimly in the west.

We made a quick breakfast on fruit like grapes and a wheaten gruel, and hastened to the chamber where we had been received the day before. Zaphnath was already there, and so were eleven of the grey-beards. We did not wait for the twelfth, but Zaphnath led the doctor to the place at the centre of their oval table, which thus filled all the seats. Then the young ruler ascended his throne and thus addressed them:—

"While ye have tossed and tumbled in an idle slumber, two things of grave importance have happened touching you. The Pharaoh, acting upon my urgent advices, hath appointed this grey-beard from the Blue Star to be your chief; and now the Blue Star herself hath re-appeared upon the very face of the Day-Giver, even as these, her people, told us yesterday that she must do."

Just at this point the belated wise man came straggling in, a slow surprise growing upon him when he saw that his seat was taken. Zaphnath then turned, addressing him,—

"Thou hast not heard, O lazy idler in the lap of morning, what I have just spoken to thy brothers? Then go thou to yonder Larger Eye and speak truthfully to these grey-beards all that thou seest."

I adjusted the instrument, and placed him in the proper position to see. He looked long and carefully, then left the instrument and looked with the unaided eye. Coming back he gazed again, and finally spoke very slowly, as if resigning his life with the words:—

"I am old, and my sight deceiveth me, O my brothers, for when I gaze into this mysterious instrument the Day-Giver suddenly groweth very large, and hath two blots of shadow upon the upper half of his brightness. But when I look with my proper eyes, he keeps his size, and there are still spots upon him, but they are upon his lower side."

I explained to Zaphnath that the telescope made things look wrong side up, just as it made them look larger, and I focussed it upon the Gnomons to convince the wise man of this. Then the youth spoke to him again:—

"The Pharaoh hath appointed this grey-beard

from the Blue Star to be chief of all the wise men, and as there can be but twelve, thou art no longer one. Unto thee, however, is given the duty of teaching our language to the chief. See that thou doest it well, for the lives of all of you, having now been forfeited by the law, are in his hands. But so long as his wisdom spares you, ye shall live."

As there was now a lull, I saw an opportunity for my plan which I had not yet found time to explain to the doctor. I translated to him as I proceeded, however,—

"Tell me, O Zaphnath, is it the custom here to relate dreams to the wise men for interpretation? I had last night a most peculiar one, and I will give this golden coin to whomsoever is able to explain its meaning." All the great eyes opened wide and round at beholding the eagle I held up to view. So large a piece of gold must have been uncommon. The youth replied,—

"It is, in truth, an obsolete formality to submit dreams to the wise men, for they have interpreted none since I came into Kem. But let us hear it; if they cannot make it known, mayhap I can do so."

"I dreamed that I stood by the great river which runneth just without thy city walls, and I saw coming up out of the water, as if they had been fishes, seven familiar beasts, such as I have not seen since I came to Kem. Knowest thou here such large, useful animals, each having a long tail and four legs, and whose peaceful habit is to eat the grass of the fields, which, having digested, the female yieldeth back in a white fluid very fit to drink?"

"It is kine thou meanest," answered Zaphnath. "In truth there are but few within the city, but they are well known, for in the land of my father my people do naught but to breed and raise them and send them hither for ploughing in the fields. At the season of planting thou shalt see many of them."

"I saw seven kine, most sleek and plump of flesh, feeding in a green meadow by the river; but suddenly there came up out of the water in the same manner two lean and shrunken kine, whose withered bones rattled against their dry skins, they were so poor and hungry. And they stayed not to eat the grass of the meadow, but fell upon and devoured their fatter sisters——"

"Saidst thou two?" interrupted Zaphnath.

"Two of the lean and shrunken, but they ate the fat-fleshed, which were seven," I answered, watching Zaphnath and the wise men closely, for he was translating to them phrase by phrase as I spoke. He faltered when I described the eating up of the fat cattle; there were wondering and inquiring looks among the wise men and a constant chattering in Kemish. I waited patiently for some time, then waving my coin I demanded,—

"Can none of the grey-beards declare the meaning to me?"

There were more consultations among themselves and with Zaphnath, and presently he said,—

"Before the wise men can declare thy dream, they demand to know whether the lean kine only slaughtered the sleek ones, or if they ate them wholly up? And were they filled and satisfied when they had eaten their fatter sisters?"

"In truth, I forgot to say that they devoured the fat kine wholly and completely, yet it could not be known that they had eaten anything, they were still so lean and ill-favoured."

This caused even a greater chattering than before, and the youth finally asked,—

"Didst thou dream aught more, or is this all?"

"Truly I had another dream, but it was different. I thought that all the wheat in the field grew upon one stalk in seven great kernels; then a shrivelled and withered stalk began to spring up; when suddenly a rapping on my door awakened me, and I dreamed no more."

The effect which this produced was most curious. Blank surprise, hidden cunning, anxious debating and uneasy hesitation, succeeded each other among the wise men. I watched it with great interest, and perceived the doctor's satisfaction, but I again demanded the interpretation.

"Know, then, O dreamer," answered Zaphnath, "that we understand not only the import of all that thou hast dreamed, but even what thou wouldst have dreamed hadst thou not been wakened! But, in spite of thy handsome offer, it doth not appear fit or proper to us that the interpretation of it should be made known to thee. Tell me, however, hast thou had conversation with any other person in Kem, save with me and with the wise men?"

"Thou knowest well, O Zaphnath, that I speak not the Kemish tongue, and can understand or communicate only through thy interpretation. I have spoken with no one on all of Ptah except through thee, and if thou wilt not declare my dream I care not, for while ye have been debating among yourselves I have learned its meaning!"

"Thou understandest it already!" he exclaimed. "Pray tell us, then, how thou hast learned it."

"The chief wise man hath declared it to me in my own tongue!" I exclaimed, with a meaning look toward the doctor, who had been speaking to me to urge caution. "He saith that the seven sleek kine are the Kemish people, and the two lean and ill-favoured are we two from the earth—for are not thy people larger and plumper than we!—and the seven denoteth their much greater number. But the dream meaneth that we two, poor and hungry, might eat up all your people and become their masters."

There was still more delighted jabbering and excited comment. Then Zaphnath arose, and turning graciously to the doctor said to him,—

"Thy marvellous interpretation, O chief greybeard, is most correct and wise, and it hath wholly eaten ours up! We quite agree with thy superior wisdom, for thou only hast read the dream aright!"

CHAPTER IX

Getting into the Corner

THE doctor's new official position carried with it the use of a spacious, rambling dwelling, situated just inside the gate where we had met Miss Blank. It was thus conveniently located for the doctor's duties at the observatories on the plateau. Another house would have been assigned to me, but I preferred to live with the doctor, and I desired to keep my eye on those enormous stone structures which our telescope had quickly relegated to scientific uselessness.

We had established ourselves comfortably in this house, surrounded ourselves with a modest retinue of servants, and were rapidly becoming acquainted with Kemish life and manners. The doctor learned the language laboriously from the deposed wise man, who had no means of communicating with him except in the tongue he was teaching. Thus it happened that the doctor could teach me in a few hours in the evening what it had taken him all day to learn. Naturally I picked up the most common phrases used in receiving and handling the grain, by hearing them frequently; but I soon learned that I must pronounce them with exactly the same intonation and emphasis, or they were not understood. Knowing but one language themselves, they had no facility in recognising mispronounced words, or in guessing at the meaning of incomplete phrases on which I stumbled.

The most difficult thing I encountered was their method of telling the time. During the day it was reckoned rationally enough by the passage of the Sun, which was never obscured by clouds and could always be seen. Every house had a small hole in the roof, at a fixed distance from the floor, and the daily track and varying shape of the spot of sunshine thus admitted gave names to the periods of the day. There seemed to be a settled superstition that no house was fortunate unless this spot of sunshine entered by the door in the morning. For this reason the principal door in nearly every house was built in the west, so that the rising Sun would cast its spot first on the porch outside and then gradually creep in through the door, across the floor, and up the opposite wall late in the afternoon. Of course there were daylight periods in the early morning and late afternoon when the Sun was too low to cast a spot, and these were known by terms which are best translated "before the clock" and "after the clock."

No one dared to make a social call while the Sun was still outside the door, but friends were best welcome when the Sun was just entering it. Moreover, whoever slept until the Sun had entered the door was looked upon as an irredeemable sluggard. The track of the spot from the door-sill to the wall opposite was measured by linear distance from the centre or noon-position of the spot. As in different houses the apertures through which the clock-light was admitted were always the same distance from the floor, such expressions as "two feet before noon," or "a foot and a quarter after noon" (which I translate from the Kemish) always had a definite and exact meaning. The nearer the spot drew to noon the more exactly circular it became and the more slowly it moved. Therefore, very fine measurements were needed in the middle of the day, and an inch near noon represented nearly as much time as a foot in the morning or evening.

But the daylight methods were simplicity itself compared with the night methods, which were calculated on an entirely different system, based on the combined movements of the two moons, neither of which agreed or coincided with the movement of the Sun in any close degree. I urged upon the doctor, as one of his earliest duties, the necessity of reforming their calendar and establish-

GETTING INTO THE CORNER

ing a uniform method of denoting the time, to extend throughout the day and night. But on this point he failed to agree with me.

"What are our seconds, minutes, hours, and weeks after all?" he queried. "They are only arbitrary and meaningless divisions of time, which we have found necessary because we have a very meagre heavenly clockwork; but here they have a very elaborate one. Our day is a rational period based on the Sun's revolution. Here they have seen fit to give up the Sun-day to simplify matters and stick to a Moon-day. Their two contrary moons furnish a rational, if rather intricate, method of telling the time at night. They are best understood by imagining them to represent the two hands of a clock. The smaller moon is what may be called a 'week hand,' completing its revolution in five and a half Sun-days; which they have for convenience divided into six Moon-days of twenty-two hours each. The larger moon makes two complete revolutions in a day, just as the hour hand of a clock does; and it really makes but little difference that it travels around the dial in an opposite direction to that of the 'week hand,' or that they both gain two hours a day on the Sun. These are mere details, that one gets used to in the end."

"Doctor, you argue like the old farmer I used to know, who stuck to the clock handed down by his grandfather, and maintained that no newfangled arrangement kept as good time. It was true that the striking apparatus had long ago failed to agree with the hands; and the hands themselves, owing to the accumulated inaccuracies of years, no longer denoted the real time; nevertheless, whenever it struck seven he could always be sure that the hands were pointing to a quarter-past twelve, and it was then just twenty-two minutes to three. This was something he could depend upon with a certainty which quite compensated for the annoyance of incessant calculations and mental corrections."

"Pray leave joking aside and consider the wonderful nightly clockwork here, which makes automatic time-keepers unnecessary. This accommodating inner moon, within the brief space of five hours, goes through the phases of a thin crescent, first quarter, and just as it approaches fulness it submits to a total eclipse, followed by a waning quarter, then the reverse crescent of an old moon, and finally it sets where the Sun must soon rise. It is a wonderful heavenly clock, which is never obscured by clouds, and turns its face toward every one alike."

"Yes, but one must remember that this hurrying moon gains two hours a day on the Sun, and therefore goes through her performance that much earlier each night. Besides, she is capable of rising twice in the same night occasionally."

"Those are mere details that one learns to allow

for. Moreover, consider the convenience of being able to tell the day of the week by the smaller moon. If it is just risen, we know we are on the eve of the first day of the week; if it is high or eclipsed, it must be the second day; and if it is sinking in the west, it is the third day——"

"But for the last half of the week it is not seen at all, and one is free to guess which day it is," I interrupted. "Then no two days of the week begin at the same hour. The first day begins with sunrise, the second two hours before sunrise, the third four hours before, and the fourth at midnight, and so on—two hours earlier each day till the week ends, when they throw in a whole night for good measure and begin the next week at sunrise again!"

"Yes, that arrangement is made necessary because their Moon-day will not agree with their Sun-day in any other manner. But it is rather remarkable that the two moons agree with each other so well, the larger one making twelve revolutions while the smaller makes one, so that at the end of every week they both rise together, but on opposite sides of the horizon, which is the signal for that night to be disregarded in the count. The next week begins on the following morning, the first rising of the larger moon being disregarded, and her second rising being the one reckoned from."

We were discussing this during our noon-day

meal, and, when we had finished, I walked with the doctor out to the plateau, where I was supervising some important work on the Gnomons; for I had not been ten days in Kem until I attempted to buy, with my gold coins, a large amount of wheat from the Pharaoh. Through the interference and objection of Zaphnath, however, I failed utterly in getting any. But the gold had its effect just the same, and later the Pharaoh showed an evident willingness to part with anything in his possession in order to get a liberal number of the smaller coins. But I put a very high value upon the gold, comparing closely with the worth of diamonds upon Earth, and refused to part with any, until one day the wisdom of buying the Gnomons occurred to me. I considered the project carefully, and finally made him an offer of a hundred half-eagles for them. Many of the small ones had been built to watch the course of the birth-stars of his various ancestors, and these were now in a sense monuments to his dynasty. He reserved these and a small one, built to observe his own star of nativity, and finally sold me all the large important ones, upon the doctor's representation that they were no longer needed for astronomical purposes. He specified only that they must not be torn down, but that I might use them as I should see fit.

As I have said before, the Gnomons contained numerous large, long chambers, and it only became necessary to put a permanent bottom in these to convert them into enormous warehouses. All the storage places inside the city were rapidly filling with grain, which poured in at every gate on tens of thousands of mules. The plenteous crop, already ripening, would have to be housed somewhere, and even if I did not succeed in buying a large store of grain for myself, I knew how to make a storehouse eat up a large portion of the value of the grain it housed. I had seen wheat, stored year after year, finally become the property of the elevator owner, by virtue of his charges.

I was not only putting a bottom to the storage chambers, but converting the inclined slopes of the largest Gnomons into a passable mule-trail, by roughening and corrugating the surface to give the patient animals a surer foot-hold, so they might climb to the top to discharge their cargoes. This was a simple form of elevator, and I laughed to think what some of my former acquaintances would think of it! One of the smaller Gnomons had already been completed to receive my share of the grain which I earned in the Pharaoh's service, and to this I was adding such meagre purchases as I could make from the small farmers. These, however, were not numerous, for the land was mostly in the hands of the Pharaoh and of a few large owners, more or less bound to him. I was therefore not a little surprised now upon approaching to see a long line of mules picking their way up the inclined side of the finished Gnomon, and as they reached the top their drivers emptied the pair of sacks they bore into my storehouse. Including the drove of unladen animals at the bottom of the Gnomon, there must have been a hundred in all, and I was awaited by the chief driver, who rode one sleek mule covered with a soft blanket of feather texture, and had another similarly saddled by his side. After a slow salute of each hand upon his cheek, he said to me, —

"My master, the glorious Hotep, sendeth to the keeper of the Pharaoh's grain a present of two hundred bags of wheat, and wisheth to know if it be true that thou desirest to buy a large store of grain with gold? For hath not Hotep the gathered harvests of two full years in his bins, and upon his fertile lands the largest crop in all Kem (save only that of the Pharaoh) is nodding and awaiting the warm, ripening breath of the Snowless Month! Yet Hotep hath no use for iron money, for he is weighted and fettered with it already; but if thou desirest to bargain with him for as much yellow gold as thou hast bartered to the Pharaoh, he will be most pleased to treat with thee, and sendeth me with this ambling mule to fetch thee. Will it please thee to come with me now to his palace within the city?"

"What do you think, Doctor? This Hotep must be almost a rival to the Pharaoh, if he has stored so much grain and owns so many ripening fields. He must have seen the new gold ornaments upon the Pharaoh's women, which have rendered him envious. If, indeed, he has such a vast quantity of grain to sell, I will deck him out with gold, such as will turn the Pharaoh green with envy! I shall lose no time in seeing him;" and so saying I mounted the mule, and assured the chief driver I would express my thanks in person to the great Hotep.

He conducted me to the opposite side of the city, and, as we crossed a height near its centre, he pointed out to me the long fields of his master lining the left bank of the river. There were miles of waving grain just beginning to turn from a luxuriant green to the lighter yellow tints of harvest. Presently we approached a large palace, which I had often before seen from afar against the distant wall of the city, but had never known. Upon entering, I observed every sign of the same idle luxury which marked the Pharaoh's dwelling, but none of that regal disdain or imperial haughtiness which separated every one but his favourite women from the immediate presence of the monarch.

I was graciously received in a large, lighted chamber, where Hotep reclined lazily upon a billowy heap of downy cushions, surrounded by many women. He only arose from his elbow to a sitting posture when I saluted him. Without saying a word to him, I approached, and, loosening my belt from about my waist, I unbuckled its mouth and poured out upon a large cushion by his side my three hundred shining golden eagles. The effect was electrical, for they were twice the size and three times as many as the coins I had given the Pharaoh. It must have seemed impossible to him that I could possess larger coins, and more of them, than he had seen upon the monarch's favourites. He was simply delighted with the mere view, and his women crowded around or ran out in haste to bring in their absent sisters to behold a marvel of riches such as Kem had never seen. But though they wondered and gloated over the sight, none of them touched a coin until I spoke.

"I pray thee, most gracious Hotep, examine all these coins, and pass them among thy women to see if they be pleased with them. Observe their regularity of form and beauty of design, and test the music they give forth when cast upon thy floor of stone. Mayhap, thou wouldst rather own all these than to be cumbered with so much grain."

Thereupon Hotep seized a heaping handful, which he poured jingling from one palm to the other, and all the women delved their pretty fingers into the shining heap and passed the coins to their admiring sisters, until not one was left upon the cushion.

"Thy Chief of Harvests hath made known to me, O Hotep, that thou still hast the full crops of two years. Wilt thou tell me how many bags of grain grow upon thy fields at a single crop?"

"Are not the number of my mules a thousand and one, and bear they not two bags each? To gather a single harvest, each faithful animal must make five trips each day for the period of an hundred days."

I had often estimated an average mule-load at five bushels, upon which basis each crop would aggregate two and a half million bushels. This seemed impossible for a single farmer, but his fields wearied the sight to follow down the left bank of the Nasr-Nil.

"If thou wilt leave all this gold with me, I will deliver by my mules to thy storehouses upon the plateau all the grain of my past two crops with which my whole palace here is cumbered."

"I fear thou holdest thy grain too dearly, and that thou knowest not the value of this gold. What is more plenteous in Kem than wheat? There be more bags of it than the stars in heaven. But this gold I bring is more than all the store of it upon Ptah before I came. Pray give it back again," I said, gathering up the few pieces which had been returned to the cushion, and glancing about among the women as if searching for the rest. They returned them slowly, but Hotep still held his handful. After a brief pause, I continued,—

"Hast thou not a fair crop growing which thou

mightest also give me, so that no other than Hotep shall receive any of these coins?"

"In truth, I have never ridden as far as my waving fields stretch down the Nasr-Nil; but one cannot sell what hath not fully ripened, for who knoweth what it may turn out to be?"

"Then I must beg thee to return my coins," I answered slowly; but, unbuckling the other end of my belt, I poured out upon another cushion the hundred magnificent double eagles which I was holding in reserve. Then, taking a particularly bright one of these, I continued,—

"But as thou hast been generous and thoughtful enough to send me a present, O Hotep, I desire to return one to thee, such as no man in Kem ever possessed before. Will it please thee to accept this disc of gold as large as the lesser moon that creeps across the sky? And with it go my wishes that Hotep's crops may always be great and plentiful."

Slowly and unwillingly the women returned the eagles to the cushion, while they stared in wonder at the heap of larger coins. Hotep filtered the handful through his fingers to the cushion, and accepted the double eagle with gladness. With his eyes fixed on the second heap he seemed to be thinking deeply and making calculations.

"The people are wont to call thee Iron Man, but I believe thou art golden!" he ruminated, and then suddenly, "For these heaps of riches, large and small, what desirest thou of all my possessions? Wilt thou have all my grain and half my land? Shall I give to thee all my fields which cannot be seen from the palace here?"

"Why should I wish thy land when I have no cattle to till it, nor mules to gather the harvest? In lieu of the land, give me only a share of what it should produce for a few years. Now give heed to the bargain I will make with thee. If thou wilt deliver to my storehouses, upon the plateau, all the gathered grain of thy past two crops, and all the grain thou shalt gather from this growing crop (save only what thou needest for seed), and half of each of the crops of the three succeeding years,—provided, however, that you assure me each year as much as thy thousand mules can carry in an hundred journeys;—then thou mayest keep all this store of gold, which is, indeed, all that both of us from the Blue Star possess."

He seemed to be revolving these terms slowly in his mind to be sure of them, and then called out to his servants,—

"Bring in spiced wine, and bid my Chief of Harvests enter! He shall be witness that Hotep agrees to this compact, and, should I die before it is fulfilled, he shall see that it is carried out to the last year. But wilt thou leave all this gold with me now, or must I wait until the harvests are delivered?"

"What Hotep promiseth me I believe, as certainly as if it were done already. I will leave the gold with thee, knowing thou wilt perform the contract in every item; but if thou failest in any year, thou shalt return to me one small gold-piece for each trip that thy thousand mules fall short of an hundred."

He agreed, and arose and recited the terms of the compact to his Chief of Harvests, charging him to carry it out, and to cause to be engraved a small stone cylinder as a permanent record of its provisions, as it was their custom to do in such cases. Then filling three goblets with rich spiced wine, he exclaimed,—

"For thy sake, O most generous youth, may the Nasr-Nil fondly nurse every harvest, and may the gentle Snowless Month ripen them in such abundance as they have never shown before! And may Hotep's mules grow old and weary bearing the plenty to thy storehouses!"

CHAPTER X

Humanity on Ptah

THE magnificent abundance of the seventh great harvest, which ripened late in the year of our arrival, attracted a multitude of both men and animals from all the out-lying countries into Kem to assist in gathering it, and many of them remained to spend their gains in the luxuries of the great city. It was an unparalleled period of prosperity and plenty; and though the rich wasted everything with a careless hand, the poor were better provided for than they had ever been.

Like an endless caravan Hotep's mules trailed across the city day by day, and emptied their cargoes into the bottomless pits of the Gnomons. And Hotep's thousand cattle tramped his threshing-floors during the long winter, and until the later nightly snows signalled the coming of a tardy spring; and yet the patient mules streamed through the city, and wore deeper paths into the sides of the Gnomons, until one by one the great chambers were filled and sealed.

Late in the spring the toiling cattle left the threshing-floors, and traversed the fields in long procession, two and two, lashed together by a bar across the horns instead of a yoke, and dragging heavy stone ploughs slowly after them to prepare the soil for a new planting. But while the whole left bank of the Nasr-Nil swarmed with Hotep's patient teams and their busy drivers, the right bank was deserted, idle, and lifeless. Every one wondered why the Pharaoh's planting was being delayed; no one knew why the Pharaoh's men and cattle were idle; and the old men shook their heads and muttered that the river would overflow its banks long before the Pharaoh's seed was in. After a while Zaphnath sent for me, and when I came before him he said,—

"The Pharaoh is sick with the plenty of the land, weary of the sight of grain-laden mules and ploughing cattle, and so cumbered about with mountains of wheat that he desireth not to plant his fields. Thou art not one to see his lands lie idle. If thou hast aught with which to tempt him, I can persuade him to let unto thee all his land and to hire unto thee all his men and mules and cattle. For hath he not acquired all his riches in seven years' harvests? and in another seven thou mayest be as rich as he."

"Mayhap, O Zaphnath, the coming seven years may not be as plenteous as the last seven have been; but, in any case, I have no more gold with which to tempt the Pharaoh, having parted with all of it in a bad bargain with Hotep, whom thou knowest, for half of his coming crops."

Thereupon he bade me remain, and sent for Hotep, and said to him,—

"Behold, have not the harvests of seven years made Pharaoh the richest man upon Ptah, so that he covets no more grain, but only things of rare beauty? And are not thy harvests reduced by half through thy compact with him from the Blue Star? Now, if thou likest to tempt the Pharaoh with an hundred of thy golden coins, and one-and-twenty of the moon-sized discs of gold such as thou wearest there, thou mayest hire his land for the next seven years, and all his men and animals for a like time, if thou wilt feed and nourish them; and then shall not both banks of the great river bring forth riches, and be burdened with the plenteous harvests of Hotep?"

"Is the Pharaoh indeed weary of rich harvests, or doth he rather itch for my gold? Yet, had I the seed to plant all his fields, I might consider the undertaking thou shewest me."

"Let not that delay thee," answered Zaphnath, "for I am sure he will gladly lend to such a man as Hotep the seed thou needest until thy next harvest be gathered."

So the matter was thus finally concluded, and I was a witness to the compact.

Then Hotep's Chief of Harvests worked early and

late to finish planting before the Month of Midnight Snows, when the Nasr-Nil usually overflows its banks and waters the harvest. But, as if to oblige a man so industrious in preparing the way for it, the great river did not rise at its customary time, and Hotep was able to finish his seeding on both banks.

The black loam along the shores parched and crumbled, and borrowed the look of the great desert; the feathers of darkness fell later and later, until they began to appear with the dawn, and yet the river failed to rise; the priests went through their perfunctory rites to placate the god of the Overflow, and made their impotent sacrifices to tempt him to bless the harvest; but Hotep saw the Snowless Month, which should have ripened his grain, dawn upon fields that were dried to seas of drifting dust and void of all vegetation. His army of men, augmented by the Pharaoh's thousands, and his ten thousand cattle and mules, all ate and waited and waited and ate, and yet there was no work for them. The following spring there was no need to plough the fields, and no seed to plant them.

When Zaphnath learned that Hotep must deliver a hundred thousand mule-cargoes of wheat to me, or forfeit a hundred gold pieces, he sent for him, and sold to him for the hundred pieces enough of the Pharaoh's grain already on the plateau to pay me, and lent him the seed to plant all the land again. But aside from this, the Pharaoh sold not a bag of wheat, and during the first year all the small stores of grain throughout Kem were consumed, and the price rose to three times its former value. Therefore, Hotep consoled himself with the thought that he could make more out of one crop after a failure than he could have made out of two crops without it, and he happily sowed his fields anew.

Before the river was due to rise the second time, the poor began to suffer from the famine. There was no employment for the thousands who had been attracted to Kem to gather the previous large harvests. Only those fortunate enough to be slaves enjoyed an assured living, and this entire class was now dependent upon Hotep, for Pharaoh supported only his women and his personal servants. Many people desired to deliver themselves into slavery, but Pharaoh would not accept any, and Hotep already had more than he could feed. During the Month of Midnight Snows the entire population of the city watched the river with apprehension, noting its slightest fluctuation. But day after day the people saw no change, and idleness fostered grumbling and discontent among them. Zaphnath and the Pharaoh were privately criticised because they did not attend or contribute to the sacrifices made to the god of Overflow; because they hoarded so much grain, and did nothing to alleviate the distress of the people.

And there were many who attributed the unusual action of the river to the presence upon Ptah of two strangers from the Blue Star.

When two fruitless months had passed without any rising of the waters, Hotep lost courage, and was obliged to proclaim that all his men and beasts must exist upon half-rations. It was then that public suffering became general. About this time I consulted with the doctor whether to press Hotep for the second delivery of a hundred thousand cargoes of wheat.

"Certainly; demand it from him," he answered, greatly to my surprise, "especially so long as it amounts to squeezing the wheat out of the Pharaoh. It is certain he will furnish the wheat in exchange for Hotep's gold, and a few coins are really nothing to him or to you either. As long as the Pharaoh covets them, make him pay well for them."

"But I expected you would advise leniency, as you have never sympathized with my wheat speculation in the least," I replied.

"I do not share your idle dream of riches, but nevertheless I want to get as much wheat into our hands as possible, especially if it comes from the Pharaoh. You do not seem to appreciate the real reason, but blindly chase after the bauble of fortune. It was the same when I first saw you in Chicago, and now you are just as impulsive and thoughtless. I have no doubt but you have already computed a hundred times how rich you are in Earthly terms and figures."

"The time for a big value has not quite come yet, but I confess I have estimated that it will run into many millions of dollars."

"Rubbish! What is the use of such childish nonsense? Even if we had our projectile to return with, you could never take any of your riches back to Earth with you!"

"And why not?" I demanded in astonishment.

"What is your fortune? It now exists in grain at an inflated famine value. You couldn't transport the grain back to Earth, and if you could, it would shrink in value and fail to pay the freight. What can you exchange it for here? For lands, for women, for slaves, none of which have any commercial value on Earth."

"But I can sell it for money!" I put in.

"Yes, for iron money worth a few dollars a ton on Earth! Why, not even your entire fortune will buy enough iron to build a new projectile to enable us to return. You parted with the only valuable and portable form of property when you exchanged your gold. Now that is rapidly going into the Pharaoh's hands, to remain there, and you can never return to Earth as rich as you left it, though you be worth all the money and property in the land of Kem!"

"Well, it does look a little as if I had been

scheming for riches here, without knowing just why I want them."

"Yes, you have formed that habit on Earth. Only they carry it further there—swindle their brothers, deceive their parents, oppress the weak, extort from the poor; work, toil, plot, cheat, rob, yes, even *kill!* in order to lay up a store of something they can never take away with them, and which renders them unhappy oftener than happy while they remain to guard it."

"I have heard that sort of talk often before, Doctor, but I never saw the truth of it quite so plainly as now. I have outwitted and squeezed Hotep, the man on whom the whole city now depends for existence."

"They think they depend upon him, but you know as well as I do that he will be powerless; that he must see them starve by thousands, and part with the last bit of his cherished riches to save his own life. No, Isidor, your business sagacity has not been in vain, for this entire people depend not on Hotep, but on *you!* You alone have the food to preserve many of them alive through a famine and a pestilence whose horrors are just beginning. Pharaoh and Zaphnath will squeeze and pinch them, and see them die, and turn it all to their own profit; but let us constitute ourselves a relief committee, you and I. Let us set these Kemish rulers an example of humanity, as we know it on Earth."

CHAPTER XI

Revolutionist and Eavesdropper

IN Kem, where agriculture was almost the only occupation, and where the ox was helpful both in planting and threshing the grain, it was quite natural that he should be revered, or at least respected as a partner in the toil, and that a strong prejudice should prevail against his being slaughtered for food. In fact, it was not the practice of the Kemish to eat any large animals, but they confined themselves to fish and small fowl for meats. Nevertheless, I urged upon Hotep the necessity of killing some of his cattle to provide food for his miserable and poorly-fed labourers. But he stubbornly refused to do so, saying his men would rather eat the flesh of mules than of cattle.

Without being pressed for it, he paid me the second hundred thousand cargoes of wheat, which he bought from the Pharaoh with gold, as he had done before. But I divided this entire quantity of grain among Hotep's labourers, which eked out their half-rations for almost a year. I stipulated that none of this grain should be used for seed, for I firmly believed it would be wasted. But Pharaoh again lent the seed for planting a third crop, insisting that the discouraged Hotep should put it in the ground, and reminding him that the only way he could get grain to pay his heavy debts was to raise a crop.

Thenocris had not been long in learning the location of our house near her favourite gate, and it was her habit to call on us every day at the time of the noon-day meal. She always carried and caressed her white rabbit, and they came to us like two dumb animals to be fed. Her tall, stately figure, traversing the city on her daily journey to our house, soon became a familiar sight; and when the people began to be oppressed by hunger, they gradually overcame their early fear of us, and followed her to our door for food. We had never turned any away, for beggary was rare enough in Kem, and no sane person ever resorted to it except in the sorest extremes of need.

Zaphnath doubtless looked with an evil eye upon the crowds that daily thronged our door to secure food. The Pharaoh rarely left his palace, and bothered little about affairs outside, and Zaphnath must have been at the bottom of an edict which was shortly issued. Nothing that I remember in Kem better illustrated the absolute power of the Pharaoh and the unrestrained enforcement of his merest whim. The edict referred to the scarcity of bread and the multitude of foreigners who were flocking to the city to secure it, and provided (ostensibly for the good of the Kemish people) that no man in the city of Kem should give bread or any sort of food to any but the members of his own household. Moreover, no man should sell grain or bread at a less price than that established by the Pharaoh for the sale of his own.

The doctor and I realized that this was aimed at no one but us. They were jealous of our charity, and wished to turn everybody's need to their own profit. We scoffed at the tyranny of such an edict, but it was the arbitrary sort of law to which the Kemish were accustomed. Yet if we gave up our undertaking, and the unfortunate multitude went unfed for a few days, bread riots were certain to break out, and they might result in the death or overthrow of the short-sighted Pharaoh, and the seizure of his grain. Even this would not settle the question, for the victors might enforce a worse monopoly of it, if that were possible.

"We must continue to feed them all outside the city,—at the Gnomons, for instance," I suggested.

"Yes, we must feed them there in a large chamber, and eat with them, so that they may be considered members of our household," added the doctor.

Thus it happened that the paths which Hotep's mules had worn so deeply were now thronged by a great multitude of the city's poor in their daily pilgrimage to the Gnomons. In an enormous chamber which we fitted up for that purpose, we served to each comer one generous meal, and there were so many who came that this meal was going on almost all day long. The Pharaoh fed no one but his favourites and his soldiers, and of these last he discharged a large number, reducing his army to a hungry, ill-fed thousand men. Those who were discharged came to eat with us, and many of those retained would gladly have done so, had we not excluded every one in the Pharaoh's service.

Meantime the Nasr-Nil ran lower in her banks than ever before, and gave no signs of rising; the nightly snows were brief and evanescent, and the rains, which had never been copious on Ptah, now ceased entirely. Every green thing gradually vanished from Kem, and Hotep's third crop rotted or lay sodden in the ground as the others had done. He knew that I had been offered the opportunity to plant the Pharaoh's fields, and that I had not only refused, but had hoarded grain. This may have led him to conclude that I knew some reason for the famine, and I was not surprised when he sought me one day at the Gnomons. He begged a strictly private interview with me, and I conducted him to a small room I had constructed by running two thin walls of porous stone from one Gnomon to another, and covering the enclosure with a flat roof.

"Dost thou know that thou hast linked together with thy slender walls the monuments of two antagonistic dynasties?" he began. "This structure to the left was built by the fifth ancestor of the present Pharaoh, in truth the first ruler of his dynasty. The structure to the right, however, is vastly older, and was built by the tenth Pharaoh of the dynasty, from which I am directly descended. My ancestors were vanquished by dint of wars, and their powers usurped by the ancestors of this same selfish Pharaoh, who hath not so good a right to rule as I."

I think I was born without a vestige of revolutionary spirit, for I have always felt a respect for the institutions that are, and an allegiance to the powers that rule. I remember the distinct shock which this utterance of Hotep's gave me. I said nothing, but he answered the surprised look on my face.

"Thou knowest well that the entire labouring population of Kem is fed by me in my fields on one side of the city; while all the poor and unfortunate are fed by you here on the other side. What man of Kem thinks of the grand palace of the Pharaoh in the midst of the city, but to curse it? What subject who knows how the Pharaoh and his favourites gorge themselves in luxurious plenty, while he nurses his hunger, but would a thousand times rather pay allegiance to those who save him from absolute starvation? And Zaphnath, in his nightly wanderings and his daily errands of espionage, thinkest thou he overhears a public grumbler who fails to curse him and his Pharaoh, and to extol the men from the Blue Star, and the unfortunate farmer, who, until now, has been able to give the people work and sustenance?"

"Doth Zaphnath spend his time in watching and spying, then?" I asked.

"Aye, that he doth! I crossed his path even now, coming through the city, and he set at following me, but by quick turns I eluded him. He it is who by his loans and compacts hath snared and tricked me until now I am utterly ruined, unless I can claim my rightful turn at ruling. Alone I cannot do it; with thy help I can."

"How, then, could I be of assistance to you?" I exclaimed in some astonishment, without stopping to think of the justice of his claims.

"From what I have heard of the thunder thou commandest, and the lightning thou art able to carry, it doth appear that thou couldst overcome the Pharaoh and his thousand half-starved men, who secretly itch to change masters. Thou hast the means to do it; I have the right to do it; and the people would unanimously applaud the doing of it. Let us strike together, then; let us seize the Pharaoh's grain and apportion it among our supporters and the needy, and when I am established as Pharaoh, thou shalt be my ruler in the place of Zaphnath."

"Thou temptest me but little, O Hotep. Once before I was offered a rulership in Kem which I refused. Besides, am I not bound by an agreement to loyalty and obedience to this Pharaoh?"

"Aye! Even as I am bound to come to a sure ruin; and as every man in Kem is bound to sit meekly by and starve. But is a ruler no way bound? May he claim the life of his subjects for his profit? How long will they suffer such treatment? And if we are restrained by loyalty, how long will it be till some one else strikes the blow we stick at——?"

He was interrupted by a vigorous knocking at the door, as of one who commands rather than entreats an opening. Who could it be? I turned to see, but Hotep caught me by the arm.

"Before thou openest, tell me if thou wilt join me in this undertaking for the sake of a suffering people?"

"Nay, Hotep; it is wrong, and I will not do it. I am bound to this Pharaoh, bad as he is, and to thy dynasty I owe nothing." The rapping began again and more loudly now, but Hotep still restrained me.

"For half of all my fields wilt thou furnish me the grain to pay the Pharaoh, and thus avert my ruin?"

"And if I would, how wouldst thou feed the men and mules and cattle through another year of famine, and another, and another?"

T

"Thou thinkest the crops will fail yet three more years!" he exclaimed, half stupefied by the thought.

"Aye, four! I know it for most certain," I answered, and the insistent knocking was vigorously renewed.

"Then I am too deep in the mire for thee or any one to pull me out. Open to this importunate knocker."

I threw open the door, and there stood the keen-eyed, angry-visaged Zaphnath! How long had he been listening outside there? How much had he stealthily overheard before he began knocking? All the Kemish had need to speak doubly loud to us from earth, for our ears were not made for thin air and its weak sounds. Moreover, Hotep had spoken throughout with a fervent declamation. But what I said in my ordinary tones was always easily understood by Hotep's keen ears. Therefore it seemed quite certain that Zaphnath had heard through the thin wall all that Hotep had said, and probably none of what I said. So much the worse. He had doubtless supplied my speeches to suit himself, and made them fit into Hotep's plotting. At any rate there was hot anger in his face when he spoke to me,—

"Thou servest the Pharaoh well, by contriving how to cross his wishes at every point! It were well thy office were withdrawn; I have brothers about me now who could better fill it."

"Whenever it pleaseth the Pharaoh or his allpotent ruler to abrogate his compact with me, I am quite ready to begin where we left off when it was made," I retorted. I did not think till afterwards that this might serve wrongly to indicate to him the tenor of my answers to Hotep's scheming. His eyes flashed angrily at this, yet he made no reply, but spoke to Hotep instead.

"Before the end of the clock this day, the Pharaoh requireth of thee full settlement of all thou owest him. Attempt nothing but a just and full repayment, O most precious Hotep, for thy every act is watched and known to us!"

CHAPTER XII

The Doctor Disappears

HOTEP saw that he was ruined, and he went to fall down before Pharaoh and beg for mercy. The monarch, not having the courage of his own hard-heartedness, answered him,—

"I desire not to deal harshly with thee, O Hotep; for thou hast struggled desperately against an unwilling soil and unpropitious seasons. But thou knowest all my affairs are in the hands of Zaphnath, without whom I do nothing. Therefore go thou before him and do even as he telleth thee."

And Hotep, having made an invoice of all his money, and slaves, and mules, and cattle, took it before Zaphnath, saying,—

"Behold, O most merciful ruler of Kem, I have threescore-and-ten of the great golden discs, and seven hundredweight of the coins of Kem wherewith to repay the Pharaoh for the seed which the seasons have stolen from me. But I have neither food for all the men, and mules, and cattle which are the Pharaoh's, nor yet for mine own; wherefore

I beg of thee to take back his slaves and animals, and release me from feeding them; and I will forfeit unto the Pharaoh all my working slaves, which are thirty score, and all my mules, which are a thousand and one, and all my cattle, which are an hundred score, and they shall be his for ever."

"Methinks thou borrowest with a large hand and repayest like a very miser," answered Zaphnath. "All the money thou namest will not buy a thousand cargoes of grain, for behold, is not wheat worth iron money, weight for weight? And to reimburse the Pharaoh for feeding all his men and animals through the famine, which may continue, it is a rare kindness in thee to desire to give him also all of thine to be fed and nourished! What wilt thou do with all thy land when thou hast no men or beasts to till it? And how wilt thou maintain thy proud palace, with three hundred women, when thou hast no revenues left?"

"'Tis true, O Zaphnath; and if the Pharaoh covet them, take them all—the palace, the women, the rich clothing and rare jewels, and even the endless fields which have cursed me! For the days of Hotep's riches are ended. Let him be acquit, and go from thee in peace!"

"Even with them all, thou knowest he is but poorly paid; yet it is I who have prevailed upon him not to be harsh with thee. But if the famine continue, what thinkest thou of doing to gain a living?"

"By my beard! Doth the Pharaoh wish to make a slave of me also?"

"Nay, Hotep; not a common slave. But hast thou a mind to starve? I have besought him to give thee an honourable and luxuriant service, befitting thy tastes and habits. He will make thee chamberlain of his palace."

"Is there no other thing thou canst think of or invent, O most merciful Zaphnath? Lands, slaves, animals, money, women, jewels, palace, and even my life and body for the gracious Pharaoh's service! Is that all? If so, I beg thee declare the bargain made and all my undertakings fully acquit."

Hotep came to me the following day, with his beard shaven and the Pharaoh's bird-wing on his brow. He wore the dress of the Pharaoh's chamberlain, and he told me how it had all happened. He also told me that the Pharaoh had now thrown wide open the doors of slavery, and offered to feed all who surrendered themselves to his service for life. And Zaphnath never ceased to itch for all the lands, and cattle, and slaves of every one in Kem and her tributary countries, either in exchange for the bare needs of life, or as pledges for seed which he knew would only rot and ruin the borrower.

I went about my affairs on the plateau that day, wondering how long I should continue there, or whether my threat had been effective in silencing the enmity of the rulers. When I returned that evening, I did not find the doctor at the house. My servant said that a messenger from the chamberlain had summoned him on important business, soon after the noon-day meal. I waited a little longer, and then I began to fear that the chamberlain had been used to decoy the doctor into some trap. If he was staying away of his own account, why did he not send me some word? Messengers were plenty. At last I sent the servant to the palace to inquire and search for him. After a long stay he returned, saying the doctor was nowhere to be found. No one had seen or heard of him there that day.

"And the chamberlain?" I demanded.

"He was not to be found in his rooms, and no one had seen him since noon-day."

"Didst thou make inquiry for the messenger who summoned the doctor?" I asked.

He had not thought of it; so I started to the palace myself. I had gone but a few steps when it occurred to me to act with a little more caution, and be prepared for some plot against myself. I turned back to the house, and had the servant remove the heap of pillows where I slept. Underneath was a loosened stone of the floor, and below it we kept the rifles, revolvers, and ammunition hidden. I carefully loaded all of them, and put all the remaining cartridges into our two old belts. I thought of strapping one of these about me, but reflected that this would have a hostile and treasonable appearance, so I contented myself with concealing one revolver in my coat, and then I carefully covered up all the rest, and had the servant pile the pillows over the stone slab again.

Then I went out and walked to the palace. Leaping the wall, I questioned every one I saw about the doctor, the chamberlain, and his messenger. No one had seen anything of them. The messenger was absent from his lodging, as well as the chamberlain. Either they were all gone somewhere secretly together, or they had all suffered a common mysterious fate. Unable to do anything more, I returned home full of apprehension.

I slept fitfully a few hours, and then I had a most realistic dream, which began among my old surroundings on Earth: the wheat pit, the closing of a turbulent session, the drive through the parks till I came suddenly in sight of the great spherical cactus design of the World in Washington Park. As I approached this, it seemed to leave its pedestal and move freely through space toward me. I seized one of its meridians, and, clinging tightly, was carried off over the park, over the lake, over seas of ice, through an ocean of sparkling light, faster and farther every moment, until presently my little globe refused to hold me longer, and repelled me through a long, giddy, awful fall which filled me with terror. But I landed in the dark chamber of a Gnomon, waist-deep in loose wheat. It seemed gradually to grow deeper about me, rose to my shoulders, to my chin; and as I looked up I saw Slater pouring in wheat in a steady stream. He meant to smother and choke me with it. Ah, if I only had a thousand, aye, ten thousand mouths to eat it, he could never do it. I could keep even with him. But it gradually rose past my mouth, past my nose; it covered my head and was smothering me. What an awful thing was too much food, after all! And then I wakened to find my head covered with pillows until I was half-choked for breath.

It was all so vivid I could not rid my mind of it. It seemed really to have happened but a moment ago. My mind was palpitating afresh with those Earthly scenes which had for years been fading out of it. What could it all mean? Then I thought of the doctor. Perhaps they were smothering him in one of the Gnomons. It seemed hardly probable, but the idea took a strange hold on me. The chambers were all full and sealed, but one; it had been opened, and wheat was daily being used out of it; none was at hand to be poured in. It was foolish to do so, but I could not rest until I had gone to the Gnomons to see. Of course I would find nothing there, but I should not be content till I had tried. At least, the night air and the gently falling feathers of darkness would restore my calmness again.

I had the precaution to take my revolver again, and after a very short walk I stood face to face with the great stone gate, barred and locked to confine all others within the city. The fact that it was fastened on the inside proved that the doctor's captors were not outside, or, at least, did not expect to return till after daylight. With a brisk jump I cleared the wall easily, and walked rapidly to the plateau. There was no sign of life there. I mounted the only unsealed Gnomon and shouted down into its cavernous depths. Of course there was no answer. I was now so wide awake it seemed to me quite silly to follow the promptings of a dream, so I began to return in a leisurely walk.

The night scene all about me, how different it was from those to which I had been accustomed on Earth! Out of a pink sky flakes of frozen dew were gently falling, starching the arid, verdureless soil with a glistening coat of evanescent white. Along the river bank, tall, slender, lightly-rooted trees reached far up into the breathless air, but there was never the movement of a bough or the rustle of a leaf, except from the flutter of birds. Jungles of spindling reeds also towered from waste marshes, in testimony to the easy struggle which vegetable sap had been able to accomplish over a weak gravity. Everything was eloquent with the reminder that I was on a different world; but yet, when I looked up at the starry heavens, they were the same. All the familiar constellations, changing their positions through the night with the same stately dignity, were there. The Pleiades, Orion, the Great Bear, with his nose constantly pointed at the Pole Star, made me feel that, at least in the heavens, I was at home! Only the colour of the night, the two little moons, and the planets looked different. Great Jupiter, king of the Martian night, whose brilliancy, if not his size, outrivalled the pale moons; Saturn, with his tilted ring, was visible to the naked eye; and you pearly blue star, just rising to announce the morning, was Earth. Earth, which I had so unwillingly left, would I ever see her again as anything but a Sun-attending star? Would I ever walk her familiar paths, and know my brother creatures there again?

With this thought came over me an unspeakable sense of loneliness, a depressing home-sickness, an aching yearning for that life, tempestuous as it had been. And how I despised the monotony and lowness of the Martian life; how I loathed the spreading misery of the famine, and the vile and dreadful pestilences which it was begetting! How could I ever endure the four more slow years of it which I confidently expected to ensue? What would I not give to leave it all and return!

I had retraced my steps, leapt the wall again, and as I approached our house was surprised to see, in the dim light of the coming morning, a figure standing guard at the doorway. He was a soldier, and on closer approach I saw that he wore a beard, which showed him to be a captain. But what surprised me far more was that he held awkwardly in his arms one of our loaded rifles. Here was certain treachery. Since he stood guard, he doubtless had soldiers within; and if they had found one firearm they must have found the others also. But how had they succeeded in finding them? A mere search never would have revealed their secret place. Some one who knew of their location must have disclosed it. Could it have been the doctor? Had they brought him back, and forced him to produce the arms?

In that case, now was my chance to liberate him. Fortunately they did not know how to use the arms they had captured, and I had one revolver with five good loads in it. With five telling shots I ought to be able to create panic enough to enable the doctor to get possession of another gun and help me rout them.

All this flashed through my mind in a twinkling, and just as I drew out my revolver the captain caught sight of me. He quickly shifted the rifle in his hands and tugged at the hammer. He knew nothing of the necessity of taking aim, or of the use of the trigger. It would only be by the merest chance if he hit me. I had half drawn the trigger, and was just correcting my aim, when a long flash of flame from the rifle startled me, and unconsciously I fired wild. By lifting the hammer of the rifle and letting it snap back, the captain had exploded one cartridge at random. But my careful aiming had now taught him a trick; I saw him attempting the same arm's-length aim with the rifle. He did it awkwardly enough, and pulled up the hammer with the other hand. It fell with a snap on the discharged cartridge. He could be relied on never to learn the trick of ejecting them and reloading with the sixteen that lay ready up the length of the barrel. Therefore, instead of firing again, I rushed at him to capture the rifle. But he was too quick for me, for thrusting it inside the house with a quick command, the other was handed out to him. I was now at such extremely close range that his awkward aim covered me; but I was quicker on the trigger than he was on the hammer, and with a cry the first Martian to suffer by gunpowder fell to the ground. I sprang for his rifle just as some one from inside snatched it away and pointed it at me again. Whoever had it, stood half behind the door and out of range. But I aimed at his fingers on the rifle barrel, and by a lucky chance I hit them, for the rifle dropped and the body staggered into full view. Another quick shot sent this fellow to the ground, but as I reached for his rifle, it was snatched away again.

Now I saw the absolute necessity of possessing myself of another firearm, for I had but one load left in the revolver. I felt little fear of their awkward aim, therefore I made bold to rush inside on the chance of seizing the first gun I could lay my hands on. At the same time I would be able to see the position of the doctor. He must be gagged, for he had made no answer to my frequent cries to him in English. Once inside, I saw that the room was full of soldiers—twenty at least. They had a prisoner, true enough, but not the doctor. It was my servant, whom they had forced to disclose the location of the arms.

The soldiers quickly blocked the door and began closing in on me. One seized me by each arm, but with a quick shake I threw them off. Then a third fellow clutched my left arm so tightly I could not loosen him. Had I taken my eyes or my revolver off the crowd in front, they would have been upon me in a body; yet with my left arm I was able slowly to turn the clinging soldier around in front of me and to bring him gradually within close range of my revolver. When he saw its gleaming muzzle, he broke from me and fled to the others.

Little did they know that I could not afford to sacrifice my remaining load to kill a single man. I must use it to capture the other revolver, for rifles were of no use at such short range. I manœuvred cautiously to keep most of the soldiers in front of me, and stealthily backed toward the door, where a soldier stood guard with the other weapon. I was reckoning on the cowardice of most of those in front of me, but I had failed to count on the men I had shot. As I now backed quickly towards the door, I suddenly felt the arms of the fallen man about my legs, and I stumbled backwards over him. In a twinkling the whole crowd was upon me, my revolver was seized, my arms were pinned to the ground, and the dying soldier clutched my legs in his last frenzy. I expected no better than to be shot immediately by a rifle held against my head, but their orders were evidently different. My arms were securely bound with rough fibrous thongs, and then they marched me to the palace just as the sun was rising.

CHAPTER XIII

The Revelation of Hotep

I WAS not a little surprised to see that they carried me to the same ante-room in the palace which I had occupied on coming to Kem. But it was now quite stripped of all furnishings, and over each door were hung large, closely-spun fabrics, which completely covered and concealed them from sight. There were but two little windows high above my head, and had I been free to leap up to them, they were too small to afford me an exit. Driven into a stone slab of the floor were two large bent-wood staples. Between these they placed several cushions, upon which they laid me.

"May it please the strong man to rest here quietly, aye! and to slumber if he feel the need, until my master, the worshipful Zaphnath, be risen?" sneered the leader in polite irony, as the soldiers, having unbound my arms, proceeded to tie each hand securely to one of the wooden rings. Then with jeers they left me, pointing the fire-arms and swords at me as they went. I heard them bar the doors on the outside and try them with a severe shake; then their footsteps receded and all was still.

As I lay on my back looking up at the vaulted stone roof, I had my first leisure to reflect on the desperate condition into which we had at last fallen. The arms, which had meant our supremacy, were in the hands of our enemies; Hotep, our only friend in the palace, had mysteriously disappeared; the doctor was taken, perhaps killed by this time; and I could hardly outlast the day, for Zaphnath would reserve but one fate for a conspirator who sought his place. How soon would he come, and how would he dispose of me? I remembered having seen the punishment for treason of a noble personage, with whom I had once eaten at the Pharaoh's table. He was confined at the bottom of a tight stone pit, and a heavy, poisonous gas was slowly poured into it. He could see it slowly fill the pit, and as it gradually rose toward his nostrils, he could feel his death gradually measured out to him by inches. When he had breathed it in a little, his face swelled a livid purple, he choked and strangled, staggered and fell beneath the murky surface to die out of sight. The terror of such a slowly creeping danger! the horror of such a repulsive death! I remember saying at the time that in his place I would have snatched a quick respite from the lingering agonies by strangling myself, or tearing my wrist open with my teeth. Now, as I thought

U

of it, I suddenly remembered my dream of being similarly smothered in the Gnomons by slowly in-pouring grain. A superstitious mind would have feared that dream foretold my fate, but I was rational enough to perceive that it must have been suggested to me by a vagrant memory of the poisoning I had seen.

As I lay thinking thus, I shifted my position a little on the pillows for better comfort, and my eyes wandered slowly from the vaulted roof to the daylight at the two little high windows. I started in terror at what I saw, but blinked my eyes to make sure I was awake, and then looked more intently. There was no dreaming this time! I saw clearly, and at both windows, a curling, purple stream of dense, noxious gas pouring down into the room! It was much heavier than the air, and trickled slowly down like the ghost of murky waters gradually filling up a great well. Then I turned to look at the floor, the stones were no longer visible, but a coat of muddy purple covered them to a depth of several inches, and the noisome gas already reached almost to the tops of my cushions! All this had trickled in within ten minutes, and twice as much more would rise and cover me completely. Then an awful but silent death would creep into my lungs, and my only friends, the common people of Kem, would never know how I had perished.

Did I try to strangle myself or tear open my wrist? I could not get hand and mouth near enough together for either of these expedients, had the stubborn instinct of self-preservation left them any place in my mind. I kicked away the cushions, which gave me a little more room to work my knees under me. Then by straining on my thongs I was able to lift my head and shoulders upright, and save my nostrils from the noxious stuff for many minutes longer. All the years of my life on Ptah I had been vain of my superior physical strength. Would it serve me now to break the thongs that bound me? I tugged, and pulled, and struggled until I cut the flesh, but they only drew tighter; yet at each effort I gained a little more length of thong.

The purple surface, on which death floated, crept up toward me. The room was gas-tight; the doors were so covered that they could not leak, and had I succeeded in breaking loose I could not have shaken their bars. To save myself, I must make a breach in the floor; I must pull up a slab and let the gaseous poison run out below. That was my only chance. I worked my knees back as nearly as possible to the edge of the slab into which the wooden staples were fastened, and threw all my weight and strength into the effort. The stone did not move. Yet I got more thong-room, and succeeded in doubling my feet under me to give more force to the next heave. I felt sure I could have lifted the weight of the stone if it were free, but struggle as I would, I could not loosen it from its wedged position. The purple poison had risen to my waist by this time, and in my violent efforts I had stirred it into billowing waves which occasionally surged almost to my nostrils. I had breathed a little which made me faint and giddy. I feared lest I should stagger and fall into it. Once my head below the surface, and I was most surely and horribly drowned!

I stood resting a second, anxiously thinking, planning in desperation and keeping my eyes always fixed on the rising purple. Suddenly, though I had given no tug, I heard the stone under me crunch at its edges, and felt it begin to rise a little at one side! What could have loosened it, when all my efforts had failed? No matter! if I could pull it away now and make a breach, I would at least gain a long respite. I tugged again and found it easy to pull the loosened stone up on one edge, till it tottered and fell over against me. Feverishly I watched the poison about me; it rose no longer; slowly it began to sink away. Thank God for so much!

Then suddenly I heard voices calling me. They seemed to come from below. Yes! It was Hotep in Kemish,—and the doctor in English! Were they confined in the cavern below, then? And had the gas been reserved for them, when it had finished its dread work with me? Horrible thought! If so, in saving myself I was only sending the sure poison to them. Where were they? I could not

THE REVELATION OF HOTEP

see down through the murky stuff; but I must warn them.

"Halloo! The gas is poisonous! Leap through, save yourselves! Climb out, or it will kill you!"

"Bear up!" I heard the doctor's voice begin, "one minute more and we——" Then there was a violent coughing, a door slammed, and the voice was barely heard—afar off—as through a wall. Had they escaped, then, to another room? I had no further time to puzzle what it meant, for another slab of my floor rose, wavered and fell over with a crash, and up through the purplish gas I could see a great round black thing rising, stretching high up into the room until its top almost touched the roof.

My God! *It was the projectile!*

When the breach in the floor was cleared, all the gas rushed down into the lower chamber. The projectile eased over on its side, and out of the rear port-hole came Hotep with a revolver and a sword. He soon had me cut loose, and then he told me how it all had happened.

He had been chamberlain but a single day when he discovered the existence of a secret subterranean chamber under the ante-room of the banquet hall. His curiosity led him to explore this, and in its darkest recess, unseen at first entrance, he found our projectile. It had been there ever since the day of its disappearance. During our interview before Zaphnath and the wise men, they had learned from us that others could not come from Earth without the projectile, and that we had left no third person in charge of it. It must have been with an order to make away with the projectile, and to secrete it in this chamber, that the third messenger had been dispatched that day. Also on my first evening in this very ante-room, I had heard Two-spot barking in the chamber below, and the servant, on hearing him too, had him hastily released, lest he should betray the hiding-place.

As soon as Hotep had found the projectile, he had sent for us, but it was the doctor alone who joined him. They two had been busy all that day and night repairing the projectile and storing it anew. In this manner the doctor had escaped the soldiers who came at daybreak to capture us both. Beyond the projectile, Hotep had discovered a secret passage leading outside the palace walls, which they could use on their errands of repairs without being observed.

All night they worked without disturbance, but early in the morning something happened to alarm them. They heard footsteps outside and a noise at the door which led to the palace. It probably meant death to be discovered there, but they extinguished their lights, entered the projectile, and closed the port-holes and lay there quite still. The door was opened, and soldiers bearing lights entered. But they made no search; they carried with them our swords, fire-arms, and the two belts of cartridges,

THE REVELATION OF HOTEP

which they deposited here, it being the natural place for their safe keeping. When they were gone, the doctor emerged and examined the revolvers and rifles, and finding that five cartridges had been discharged, he knew there had been a struggle with me in which I had been worsted. This caused them to hasten their efforts and make an escape with the projectile as soon as possible. All the supplies necessary to the batteries had been found intact in their places, and the compressing of air with the repaired pump and the further storing of food could be postponed till they were more free to do it.

At last the projectile lifted and worked; slowly it loosened the stones of my floor above them; but when one stone was pushed aside they noticed that the daylight did not come through the breach as it ought. They had heard my cries, and as the gas came down on them, the doctor had slammed the front port-hole, which was never wide open, and had thus saved himself. Hotep was safely shut into the other compartment with the fire-arms and ammunition.

The doctor now came down to the rear port-hole to speak to me.

"My plan is to escape now to the Gnomons, where we will leave Hotep in possession with most of our fire-arms. You can give him some instructions how to use them, so that he may defend himself. There we can finish our stores of air and food." To this I assented, and said to Hotep,—

"The Gnomons I give to thee, and all the land round about them, as a reward for thy most valuable assistance. Also I put into thy charge all my stores of wheat, to be distributed among the needy. Thou must husband them to last yet four years more, and for thine own thou mayest keep one measure in twenty. Take thou also a sword, a rifle, a revolver, and a belt of cartridges. Mayhap, to thy right to rule they may add the power to be a Pharaoh!"

I was interrupted by a noise below, as of some one opening the door of the secret chamber. All the deadly gas lurked in that room now, and it was certain death to whoever opened and entered! Yet if an alarm had been raised it was there they would immediately go for the fire-arms. I listened and heard faintly a voice of command, like that of Zaphnath, saying, "Haste, get me the thunderers!" Then, as the door below creaked open, I heard it louder: "The thunderers!" Next I heard many men in violent fits of coughing; I heard some groan and fall; but who or how many died by the purplish poison intended for me, I never knew.

It was but a moment later that hurried footsteps in the banquet-hall were heard approaching the veiled door-way. I took the revolver from Hotep, and motioned him inside the projectile. How cautiously they opened the door I could not see, for it was behind the great curtain. Presently, however, the captain who had bound me and bade me wait, drew aside the curtain, and the Pharaoh stood in the door, and behind him were a crowd of soldiers armed with cross-bows. In all the number I did not see the face of Zaphnath. They beheld me alone, and had no reason to suspect the presence of the others inside the projectile.

"Guard both the doors!" the captain commanded, and a detachment of soldiers barred the other door, as if thus to prevent me from escaping with the projectile; for of course they had not seen it rise through the floor.

"Seize and bind yon traitor!" cried the Pharaoh; "and he who hesitates shall be flayed!"

"And he who attempts it, shall die ere his first step be taken!" I replied, levelling the revolver. The captain started for me and I shot him down.

"If a man of you moves till I have entered this thing, I will kill the Pharaoh, as I have killed this dog! Ye serve him best who stand still as ye are!" So saying, I covered the trembling monarch with the revolver, and with my other hand I opened the rear port-hole; then stooping, I sprang inside with a quick motion. When the Pharaoh had recovered from his fright, I heard him cry out,—

"Cast that black thing, and the traitor inside it, into yon poisonous hole again!"

The soldiers did not fear to act this time, and the whole company seized the projectile and carried it toward the breach in the floor. As they lifted it on end to thrust into the hole, I called out to the doctor, who turned in two batteries, and gently we lifted out of their dumb hands and rose steadily till we touched the roof. There the vaulted stonework stopped us, and an exultant shout went up from below. Suddenly a score of arrows twanged against my window, but the doctor turned in two more batteries and then gradually we lifted the key of the great stone arch, broke through the roof, and the whole universe was an open sea before us!

Crouching by me at the port-hole, Hotep watched the roof collapse and tumble in. "For thy sake," I said to him, "I hope a falling stone may have crushed him!"

* * * * *

Thus ended our other-world life. In a time of activity it would never have occurred to me to write down these events. It was to relieve the uneventful quiet of our trip back to Earth that I undertook to set down all our Martian experiences in their proper order. No doubt it was the changeless monotony of that return journey which made the record appear to me novel, unusual, and at times exciting. But now, six little months again on Earth have made the more than three Martian years (equalling six years of Earth) seem slow, tame, and profitless. If they were pregnant with adventure, they lacked the real experiences of life which have been crowded into the half-year since our return.

The very day I reached my old home I found another wheat corner more wide-spread, if less complete and impregnable, and I set to work to break it down. Thus the maelström of modern commercial life dragged me into its dizzy whirl before I slept the first night on Earth, and I am already surfeited with it. I seem to take the Earthly life in too large and rapid doses. Into the half-year she has put a flattering success and a dismaying failure. She has given me a month of her sweetest experiences and another of her bitterest disappointments. As if she knew I would not remain long at her feast, she has served to me in quick succession a measure of renown, a taste of fortune, the rapture of wooing, the bliss of marriage, and the rare delight of loving a soul created to love me. Then one little drop from the cup of Death embittered the whole feast and turned me against it all.

In the rush and turmoil of it all I should never have thought of my crudely written narrative again had not my cousin Ruth, who never tired of the story, fished it out and sent it to a literary friend in Boston. It was probably the instant success in the scientific world of Dr. Anderwelt's scholarly books on *Mars and His Life*, and the new direction given to modern thought by his *Theory of Parallel Planetary Life*, which led my literary sponsor to think the world would be interested in a plain, unscientific narrative of our trip Marsward and our doings there. In agreeing to look it over and cause it to be a " good delivery " in the literary world, he exacted a promise from me to make my recent Earthly experiences and our adventures on Venus join in producing another story. For before the eyes of the first reader have reached these words, Dr. Anderwelt and I will have departed sunwards, on the visit to our brilliant sister planet, where, according to his theory, life will have run through some 31,000 years more than Earth toward the perfect existence. By the first return of the projectile I have promised to send back a thorough account of the evolution of life and the advancement of civilization on Venus, so far as Earthly eyes and wits can see and know it.

www.ingramcontent.com/pod-product-compliance
Lightning Source LLC
Chambersburg PA
CBHW022044230426
43672CB00008B/1060